# DEFENDING HYPATIA

# Archimedes

NEW STUDIES IN THE HISTORY AND PHILOSOPHY OF SCIENCE AND TECHNOLOGY

VOLUME 25



*Archimedes* has three fundamental goals; to further the integration of the histories of science and technology with one another: to investigate the technical, social and practical histories of specific developments in science and technology; and finally, where possible and desirable, to bring the histories of science and technology into closer contact with the philosophy of science. To these ends, each volume will have its own theme and title and will be planned by one or more members of the Advisory Board in consultation with the editor. Although the volumes have specific themes, the series itself will not be limited to one or even to a few particular areas. Its subjects include any of the sciences, ranging from biology through physics, all aspects of technology, broadly construed, as well as historically-engaged philosophy of science or technology. Taken as a whole, *Archimedes* will be of interest to historians, philosophers, and scientists, as well as to those in business and industry who seek to understand how science and industry have come to be so strongly linked.

For other titles published in this series, go to
www.springer.com/series/5644

Robert Goulding

# Defending Hypatia

## Ramus, Savile, and the Renaissance Rediscovery of Mathematical History

Springer

Prof. Robert Goulding
University of Notre Dame
Program of Liberal Studies
313 Decio Faculty Hall
Notre Dame IN 46556
USA
Robert.D.Goulding.2@nd.edu

ISSN 1385-0180
ISBN 978-90-481-3541-7 e-ISBN 978-90-481-3542-4
DOI 10.1007/978-90-481-3542-4
Springer Dordrecht Heidelberg London New York

Library of Congress Control Number: 2009941801



*Cover design*: Boekhorst Design b.v.

Printed on acid-free paper

Springer is part of Springer Science+Business Media (www.springer.com)

# Acknowledgements

I could not have pursued this project without the extraordinary resources and helpful staff of libraries in both the United States and Europe. I completed much of the manuscript and early printed book research in the British Library, the Bodleian Library, and the Vatican Library. I also consulted volumes in the University of Chicago's Special Collections Research Center and in the Special Collections section of the library of the University of Notre Dame. The Lilly Library of Indiana University and the Bibliothèque Nationale were both generous and prompt in providing crucial microfilms of rare texts. I spent a few very enjoyable days at the library of Saint Louis University, studying Ramus's own working copy of his *Scholae mathematicae*, and examining material from Walter Ong's personal archive; I also used the resources of the Vatican Film Library in St. Louis. Most of the research on the controversy between Ramus and Charpentier was completed at Harvard University's Houghton Library. The microfilm department of Firestone Library, Princeton University, was another excellent resource. Finally, the Interlibrary Loan division of the University of Notre Dame Library went to great lengths to find almost every text, both microfilm and printed, that I needed while writing up the book. I am very grateful to the staff of all these libraries.

For generous financial support I am grateful to the British Council, whose Commonwealth Scholarship brought me from New Zealand to the Warburg Institute; St. Hugh's College, Oxford, where I spent 3 years as Elizabeth Wordsworth Junior Research Fellow; the National Maritime Museum, Greenwich, for a year's support as a Caird Research Fellow; and the Princeton Society of Fellows, in whose happy intellectual environment I spent a very stimulating 3 years as a Cotsen Fellow in the Liberal Arts. In the final phase of research, I was supported by a Faculty Fellowship from the National Endowment for the Humanities, which I spent in the idyllic setting of the American Academy in Rome. I am also grateful for material support from the Bibliographical Society (UK), the Humanities Council of Princeton University and Notre Dame's Institute for Scholarship in the Liberal Arts.

The first seed of this project was planted by Anthony Grafton. In a visit to the Warburg Institute when I was an M.A. student, he suggested that I take a look at the lecture notes of Sir Henry Savile, in the Bodleian Library. Over the years, as I delved ever deeper into these documents, Tony has provided helpful advice, close readings, and a lively sounding board. His initial suggestion led to an M.A. dissertation in which I first attempted a survey of Savile's lectures, and to a Ph.D. dissertation entirely devoted to Savile's various works. Both were supervised by Jill Kraye, whose exacting standards of scholarship and insistence on clarity have been the single greatest influence on me as a scholar. I cannot thank her enough for her close attention to my words and her infinite patience with my working habits. I am also grateful to Jill and to Martin Davies for warm hospitality in London and frequent doses of great Nepalese food.

The Warburg Institute is a legendary institution filled with remarkable people, and I owe something to everyone I shared those walls with. Charles Burnett served as a co-director, and then and afterwards often provided valuable criticisms and advice. I would also like to thank Jonathan Woolfson, who gave me an early and valuable opportunity to present my work on Savile to English intellectual historians; and Peter Mack, whose course on Renaissance rhetoric and dialectic first introduced me to Ramus. Ian Jones was a constant source of encouragement and caffeine. My external examining committee, Nick Jardine and Stephen Clucas, offered useful observations and criticisms.

I owe a great debt of gratitude to Mordechai Feingold. He has read and commented on much of my work on Savile and Ramus, and has provided frequent encouragement and guidance to my research. He was also instrumental in the publication of this book, by introducing the manuscript to Jed Buchwald. Jed has been an exemplary editor; I am very grateful for his directness and honest appraisal of my work.

Over the years, many others have offered advice and useful direction, particularly in response to parts of this book that were presented as papers. I would like to thank especially the late John Fauvel, Judith Field, Owen Gingrich, Roger Highfield, Kristian Jensen, Noel Swerdlow, the late Tom Whiteside, Jim Bennett, Nick Popper, James Byrne, Lauren Kassel, Jerry Passanante, Kristine Haugen, Deborah Harkness and Pamela Smith. Eileen Reeves has been a constant friend since we first shared optical interests at Princeton, and I am grateful to her for her unfailing enthusiasm and support.

At Notre Dame, I have benefitted from a lively and supportive intellectual environment. The Program of Liberal Studies is an extraordinary, interdisciplinary department, where I have learnt more than I have taught. I have been particularly fortunate in finding there two other historians of science: Phil Sloan has been a supportive mentor, and Tom Stapleford a very amiable colleague and teaching collaborator. I thank them and all my colleagues in the Program. Notre Dame's Program in History and Philosophy of Science has been a second home at the university; I thank especially its director, Don Howard, for his good advice. Last, an informal writing group of assistant professors at Notre Dame read several chapters of the book with meticulous care: thanks to Emily Osborn, Wendy Arons, Sara Maurer and Sophie White.

Portions of this book were presented at conferences and colloquia. In particular, conferences at the Warburg Institute, Birkbeck College and the William Andrews Clark Memorial Library were important testing grounds for this material. I also aired this research at meetings of the History of Science Society and the

Renaissance Society of America; comments from other participants (some of whom, unfortunately, I do not know by name) often significantly affected my interpretation of this material. So too did comments and critiques from Notre Dame's own Intellectual History Seminar. I am grateful to all of these communities of learning.

My wife, Margaret Meserve, is not only my partner in life but my model of a great scholar. She has read and unsparingly critiqued everything I have written. For that, and for so much more, I am ever grateful. Our children Alice, Tommy and Lily remind me constantly of what is really important. Canine friend Archie was present at the beginning of this research but did not live to see its completion; his worthy successor Ramus has been a better companion than his namesake.

# Contents

# Introduction

In 1713, Pierre Rémond de Montmort wrote to the mathematician Nicolas Bernoulli:

> It would be desirable if someone wanted to take the trouble to instruct how and in what order the discoveries in mathematics have come about ... The histories of painting, of music, of medicine have been written. A good history of mathematics, especially of geometry, would be a much more interesting and useful work ... Such a work, if done well, could be regarded to some extent as a history of the human mind, since it is in this science, more than in anything else, that man makes known that gift of intelligence that God has given him to rise above all other creatures.[1]

A half-century later, Jean-Etienne Montucla provided such an account in his *Histoire des mathématiques* (first printed in 1758, and reissued in a greatly expanded form in 1799).[2] Montucla's great work is generally acknowledged as the first genuine history of mathematics. According to modern historians, previous attempts at such a history had amounted to little more than collections of anecdotes, biographies or exhaustive bibliographies: "jumbles of names, dates and titles," as one writer in the *Dictionary of Scientific Biography* characterizes them.[3] Montucla, in contrast, was thoroughly animated by the Enlightenment project expressed in de Montmort's letter. In his *Histoire* he set out to provide a philosophical history of the "development of the human mind," as he himself described it.[4] It was precisely Montucla's vision of what mathematics *meant* and his conviction that mathematics itself must undergo change through time in order to reflect the historical elevation of the human intellect, that allowed him to transform the scattered dates and anecdotes of his predecessors into a genuine history. It is arguable that all subsequent histories of mathematics – until the most recent social histories – have been little more than "footnotes to Montucla."

In this book I argue that there were indeed histories of mathematics before Montucla which are worthy of scholarly attention. To dismiss Renaissance histories

[1] Quoted at Peiffer (2002, p. 6).

[2] On Montucla, see Swerdlow (1993b).

[3] Cited at Scriba (2002, p. 112). Note also the brief treatment and dismissal of Renaissance histories at p. 110.

[4] Peiffer (2002, p. 10).

of science as nothing more than a "cloud of fine adjectives and metaphors"[5] is unfair (although it contains a germ of truth). Renaissance histories of mathematics too had their organizing principles; and many of them were grounded in philosophical convictions as deeply held as Montucla's. As Anthony Grafton has argued with respect to the historical writings of Cardano, Rheticus and Kepler, the purpose of Renaissance histories of the mathematical disciplines was not so much to trace what actually happened, as to justify the study of subjects often derided – by humanist and scholastic writers alike – as obscure, useless and undignified.[6] But Renaissance authors did not compose their histories only to persuade others. They also constructed their historical narratives as spaces in which to think about their discipline, to define its parts, distinguish among its acceptable and unacceptable forms, and prescribe its content and method of teaching. By placing their discipline into a historical context shared by other, more mainstream humanistic arts, moreover, mathematicians could avail themselves of the large, narrative structures which Renaissance humanists had developed to account for human intellectual and cultural development, origins, progress, and decline.

Many Renaissance humanists contributed to the historiography of mathematics. The best known are the Urbinate scholars Federico Commandino, who prefaced many of his editions of mathematical works with historical essays, and Bernardino Baldi, whose biographies of ancient and modern mathematicians written in the late 1580s or early 1590s were the most extensive of the period.[7] I will, however, say little about these authors. This book is intended to be a contribution to the institutional history of mathematics, as much as it is a study of Renaissance historiography of mathematics. I will thus focus on Peter Ramus, in part because his importance as a historian has been quite overlooked (even though the better-studied Baldi and Commandino drew upon his work for their own histories), and because of his importance as a university reformer. The other principal focus of this book is Henry Savile, who offered the most comprehensive response to Ramus's mathematical writings, and held quite different views of reform at Oxford. Much of the material on which my argument is founded has been little studied: manuscript lecture notes, for instance, are discussed in detail here for the first time, and many of the printed texts (such as histories prefaced to mathematical and astronomical texts) have received little scholarly attention before.

In the Middle Ages, mathematicians had expressed little interest in the history of their discipline; indeed it was not at all evident why mathematics, the purest of sciences, should even *have* a history, or why that history should matter. In the Renaissance, by contrast, the history of mathematics flourished, even as mathematics itself lost favor with many literary humanists, to whom the subject reeked of the medieval schoolroom. Indeed, mathematics was a subject of little utility

[5] Grafton (1997, p. 262).

[6] *Ibid.* See also Goulding (2006a).

[7] Parts of Baldi's monumental work, in Italian, have been published in various places. See Rose (1975, ch. 11); and, more recently, Federici Vescovini (1998).

in this period – or so, at least, it was thought by university students in pursuit of a humanist education, in which rhetoric, literature and history would be central. Neither students nor university administrators were particularly concerned to maintain the mandatory teaching of mathematics in the university arts curriculum.

And yet, despite this de facto and official neglect, there were many enthusiastic mathematicians in the academy. The rise of the history of mathematics was, in part, a result of their concern to represent mathematics to their audience as a discipline of legitimate humanist interest. One obvious goal was simply to remind students (and fellow masters) that mathematics was not a medieval invention, but ancient; some humanist mathematicians argued, in fact, that it was the *most* ancient of sciences. Another goal was to establish the philosophical value of mathematical study. Histories of mathematics often rode the coattails of resurgent Renaissance Platonism: even if their students and readers had little interest in actually learning the subject per se, they were interested in the *idea* of mathematics, as Plato had presented it in the *Republic* and *Timaeus*. Through the history of mathematics, those who would never master the discipline itself could feel as though they had grasped its spirit, at least.

The genre of history also allowed teachers of mathematics – whether lecturing or writing for print – to do the sorts of things humanists were expected to do: collect and criticize ancient texts, harmonize readings, establish biographies and produce a rhetorically powerful, morally edifying historical narrative. These activities, in themselves, were accepted sources of cultural capital, and university mathematicians often crafted and exploited their histories as a vehicle for humanist self-promotion. Mathematics did not carry the cachet of rhetoric, history or even natural philosophy; but by emphasizing the long history of the discipline, and the literary and philological tools needed to uncover that history, its present-day practitioners could stake a claim to the same status as their humanist colleagues.

In general, the "facts" themselves were not in question. Most authors from this period agreed on a few fixed points: that, for instance, the earliest Hebrew patriarchs had an excellent knowledge of the arts; that they had preserved their discoveries from the Flood by inscribing them on stone and brick pillars; and that Abraham, "planter of mathematics" (as Gabriel Harvey marginally honored him)[8] played an important role in transmitting the sciences to other cultures. In hunting down the *prisca scientia*, their beliefs were underlined by the testimony of the Jewish historian Josephus, as well as fabulous accounts of ancient barbarian knowledge found in Diodorus Siculus and other ancient Greek historians. They may have molded their narratives to make their larger points, but there can be no doubt that most trusted the broadly-agreed accounts of origins. Claims to antiquity were an essential element in demonstrating the legitimacy and dignity of any science. But such claims also placed the historian-practitioner of mathematics in an awkward situation. While it might be impressive to discover the sciences being practiced in the Garden of Eden, even taught to the first human beings by God Himself, what room was then left either

[8] See Popper (2006, p. 100).

for individual accomplishment or (and this was most important) the extraordinary achievements of the Greeks?

The early-modern historian of mathematics and the sciences generally had to intervene in the narrative somewhere. For reasons that will become clear in the following chapters, almost all rejected a static conception of the arts. A few (like Johannes Regiomontanus and Henry Savile) embraced a narrative of progress, in which the Greeks surpassed the biblical or mythical ancients, and moderns might hope to do the same. Others (such as Peter Ramus) constructed a cycle of degeneration and (partial) recovery. But whichever model he adopted, each used his history to address the current state of his discipline. Mathematics had been a part of the medieval arts curriculum; most of the authors considered in this book were, in one way or another, concerned with the reform of the Renaissance university syllabus and the introduction of newly discovered texts or techniques into the schools. The mismatch between their ideals and the actual condition of the sciences in the academy all but demanded a historiographical model of change, whether for better or worse.

Given their presuppositions, humanist historians of mathematics understandably focused on the problem of recovery, rather than invention – a narrative which might seem at odds with the progressive optimism of the authors of the "new science," from Bacon and Descartes to Boyle.[9] Nevertheless, Renaissance historians uncovered a great deal of information on the mathematical practitioners and mathematical practice of antiquity, little of which was made explicit in the ancient sources they drew upon. They also legitimized expertise (another focus of modern histories of early-modern science)[10] for the wider intellectual culture, a crucial step in the epistemological transformations which led to the Scientific Revolution. In examining just how they did this, this book advances the debate over the role of humanism in the "scientific revolution," and the place of the sciences at the early-modern universities; it also contributes to the study of Renaissance humanism itself.

The two central figures of the book are Henry Savile and Peter Ramus. English and American scholars have only quite recently begun to take Ramus seriously again as an intellectual presence in the sixteenth century, after the devastating attacks on his reputation by his intellectual biographer Walter Ong. I examine Ramus not only as a historian, but as a philosopher of mathematics, at times even as a philosopher of mind.

## Structure of the Book

In the first part of the book (first four chapters) I examine the general development of historical narratives of mathematics, especially accounts of the origin of mathematics and its transmission from culture to culture, by authors working in the period

[9] For this point, see particularly the conclusion to Popper (2006).

[10] See, most recently, Ash (2004).

1460–1620. After an introductory survey of early Renaissance histories of mathematics and their classical sources, I turn to the central figures of this book: the French educational reformer and philosopher Peter Ramus, and the English mathematician and humanist Henry Savile. Through a close reading of their historical prefaces, letters, lectures and polemical pamphlets, I illuminate how Ramus, Savile and others fashioned their narratives of the history of mathematics, and what personal and disciplinary goals these narratives served. In the second part of the book (fifth and sixth chapters) I bring the broad debates of the first part into tighter focus. Here I investigate a nexus of historical problems which arose around the *Elements* of geometry and its author: who was Euclid, when and where did he live, what connection did he have with the philosophical schools of ancient Athens, and how did he compile the *Elements*. Throughout, I consider the impact of history (or rather, the cultivation of a historical sensibility) on the teaching and practice of mathematics in the university curriculum.

In first chapter I survey the central ancient and early Renaissance texts on which humanist historians of mathematics would rely: in particular, Proclus's *Commentary* on Euclid, Diodorus Siculus's *Historical Library* and Josephus's *Jewish Antiquities*. I then examine early attempts to craft a mathematical narrative by Johannes Regiomontanus, Girolamo Cardano and others, and I trace the emergence of a consensus around a Platonizing account that united various conflicting narratives.

The role of Proclus in this process was crucial, since he provided a historiographical model which allowed for progress in mathematics and emphasized its *philosophical* content. In fact, Proclus wove philosophy and history together so tightly that it would be difficult for later scholars – even modern scholars – to disentangle them. Renaissance narratives based on Proclus emphasized the other-worldly, contemplative nature of mathematics, portraying it as a vehicle for the transmission of perennial, divine knowledge from culture to culture. To students at the humanist university, this narrative represented mathematics as a means to other, clearly desirable ends, rather than an end in itself, and also located it within existing narratives of pagan and Christian wisdom.

Humanists also inherited from Proclus (and, it must be said, from the medieval classroom as well) the notion that Euclid represented in some way the *culmination* of ancient mathematics. The focus on Euclid (at this stage, at least, in preference to Archimedes, who would be the choice of modern historians)[11] had far-reaching effects for the whole history of mathematics. Since the development of mathematics was directed towards Euclid, humanists focused on those aspects of mathematics that they thought relevant to appreciating the achievement of the *Elements*. In particular, they emphasized the discovery of elementary theorems, generalization, ordering and classification, the synthetic method, and (following Proclus's lead) Platonism. On the other hand, they tended to neglect the solution of problems, the

[11] The Renaissance debate over shift from Euclid to Archimedes as the paradigmatic mathematician is explored in Høyrup (1992).

development of mathematical techniques, applied mathematics, and the method of analysis.[12]

Application, technique and analysis were central to Peter Ramus's reforms of the liberal arts. Interested in mathematics from early in his career, it was almost inevitable that the French logician would eventually engage with the history of mathematics, challenging and destabilizing the traditional narrative. Ramus was the single most important Renaissance historian of mathematics; the most prolific of all authors in the genre, his several works had a catalyzing effect on his contemporaries, and continued to be read long after his death. In second chapter, I trace the development of his early ideas about mathematics, beginning with the first edition of his dialectic in 1543. Ramus's reform of dialectic was predicated on the notion that an original, "natural" art had been effaced by inept and self-serving embellishments – that logic, in other words, had been shaped by history. In contrast, mathematics was outside of history. Retaining the simplicity it had from the beginning, it was the only truly natural art, and Euclid's *Elements* could be read as an authentic record of that art. That, at least, was Ramus's position for much of his early career. But as mathematics became ever more central to the Ramist program, Ramus himself became less and less enamored of Greek mathematics, and the *Elements* in particular. Over time he came to believe that mathematics was just as defaced by abstraction and inutility as scholastic dialectic; indeed, its dependence on proof and demonstration showed beyond doubt that it had strayed from the intuitive immediacy of the original natural art. A proper history would show where and how the discipline had gone astray, and how utility and self-evidence could be restored to their proper places in a natural mathematics. In a series of prefaces to his textbooks of mathematics, Ramus constructed an increasingly critical history of mathematics, a story of intellectual decline and obfuscation, which he deployed in order to reject the Platonizing consensus of earlier writers.

Ramus's work on the history of mathematics entered a new phase with his 1567 *Prooemium mathematicum*, an exhortation to the study of mathematics in which he traced the history of the discipline from the earliest times to his own day. Addressed to the Queen Mother, Catherine de Médicis, the *Prooemium* was issued both as an explicit request for royal patronage of mathematics, and as a polemical attack against his principal rival, Jacques Charpentier, who had actually managed to secure the regius professorship of mathematics in Paris despite his professed hostility to the subject. In 1569, Ramus expanded the three-book *Prooemium* into the 31 book *Scholae mathematicae*, adding among other things a proposition-by-proposition critique of the *Elements*. This was the most influential work on the history of mathematics until the publication of Montucla's *Histoire* some 230 years later; it continued to be read in the original, and digested in countless prefaces and introductions to the sciences for the more general reader. In third chapter, I examine Ramus's historical narrative in the context of his late thoughts on the philosophy

[12] All, it may be noted, central to the Wilbur Knorr's revisionist accounts of Greek mathematics, particularly in Knorr (1986).

of mathematics, and ongoing controversies over the institutional teaching of the subject.

Ramus's narrative, like so many others of the period, was built upon the idea of a *translatio doctrinae*, a transmission of learning from culture to culture, beginning with the Biblical patriarchs and ending in modern Europe. But Ramus reshaped this narrative in unexpected ways. After examining his account of the earliest mathematics, I look at the role that Pythagoras and Plato took in the later transmission of the art. Important in any account of mathematics, they appear in quite unfamiliar guise in Ramus's history. The mystical Pythagoras is transformed into a Ramist schoolmaster and author of textbooks. Plato, when he is being praised, can also look surprisingly (or not) like Ramus himself. But he emerges from Ramus's narrative as a more complex figure than Pythagoras. Although Ramus praises Plato as the "prince of mathematicians," both for his own accomplishments and for his influence on later generations, nonetheless Ramus casts an equal amount of blame in Plato's direction: more than any other, Plato started the retreat of mathematics from the popular and useful, elevating it to a level of abstraction accessible only to the philosophical elite. In Ramus's polemical account, Euclid's *Elements* represent the culmination not of mathematical accomplishment but of haughty obscurantism. Ramus's goal was to shift the attention of mathematicians away from Euclid to more worthy figures, practitioners like Archimedes and Heron of Alexandria, in whom the genuine spirit of mathematics still breathed.

In the fourth chapter I turn to examine the reception of Ramus's *Prooemium mathematicum*. The most extensive response is found in the 1570 lectures on Ptolemy delivered by the young Oxford master Henry Savile, extant in manuscript in the Bodleian Library. Savile's own mathematical self-education was founded on his reading of Ramus's *Prooemium*. He obtained a copy on its first printing, in 1567; his first forays in mathematics date from roughly the same period. Savile prefaced his close analysis and commentary of the *Almagest* with some seventy manuscript folios on the nature and history of mathematics, in which he created a narrative stretching from Adam to Ptolemy himself. He designated this preface "Prooemium mathematicum," in obvious imitation of Ramus's masterpiece.

Close examination of Savile's historical narrative together with his research notes, also extant in manuscript, shows that the influence of Ramus ran deep: the 1567 version of the *Prooemium* was the single most important source for Savile's history, and the degree of his dependence verges in some places on plagiarism. But Savile mentioned Ramus explicitly only to *criticize* him, and to defend what he saw as the authentic Hebrew and Platonic tradition of ancient mathematics from Ramus's attacks. Although he took most of his historical information from Ramus, Savile reshaped it to defend and promote a vision of mathematics entirely opposed to that of the French logician. I examine the narrative and historiographical strategies adopted by Savile, who was struggling to establish his reputation in a very different institutional setting from the Collège Royale, where Ramus was by now a senior member.

Savile's lectures were not published, although he revisited much of the same historical material in his *Praelectiones* on Euclid, published some 50 years later. More importantly, the ideals (historical and mathematical) of his lectures were enshrined in the Savilian Professorships that he founded in 1622, and which were to play such an important role in the development of mathematics at Oxford.

In fifth chapter I shift focus to a series of related case studies in Renaissance historiography of mathematics. The first concerns the biography of Euclid. The figure of Euclid played a central role in Renaissance histories of mathematics. For all historians, he represented the moment of transition from an imagined prehistory (whether patriarchal or Egyptian) to a concrete, textual history populated by Greek authorities. For some, his career also marked the high point of mathematical achievement; although Ramus certainly did not share this view. Yet the narratives surrounding the identity of Euclid were rife with errors, many of which would seem to us to have been easily avoidable. I examine how these mistakes arose and why they continued to propagate.

Erroneous narratives such as these were eventually fruitful. Far from constituting some kind of historiographical blind alley, they made it possible for philologists and mathematicians alike to treat the *Elements* as a historical text, rather than a repository of eternal, extra-historical truths, and to begin to consider Greek mathematics itself as conditioned and *limited* by historical contingencies. Modern mathematicians, conscious of those restrictions, would be able to surpass their predecessors.

The identity of Euclid was not the only controversy that surrounded the *Elements*. The text had been re-edited in the fourth century by the mathematician Theon of Alexandria. Although scholars still disagree about the precise nature of Theon's editorial interventions, most concur that his changes were fairly minor and mostly cosmetic. This is not, however, how Renaissance historians of mathematics saw it. With very few exceptions, humanist mathematicians believed that Euclid wrote only the statements of the propositions in the *Elements*; the demonstrations, they thought, were the work of Theon, writing several hundred years later. Sixth chapter surveys not only the (mistaken) historical basis for this belief, but its consequences, particularly in the editing, printing and teaching of Euclid.

Relying on complicated chronological and textual arguments, Ramus claimed to have discovered a more primitive, pre-Theonine version of the *Elements*. Although he was mistaken in his belief that he had found such a text, it enabled him to pursue an "internalist" critique of the text in the appended critical books of his 1569 *Scholae*. Ramus explained infelicities in the mathematics of the *Elements* as historical artefacts. This was the first time anyone had attempted such a philological, even source-critical study of a mathematical text; Ramus's efforts, although founded on incorrect hypotheses and conducted with the intention of not illuminating but destroying the *Elements*, anticipated those of modern Euclidean critics like Heiberg and Knorr.

Savile opposed Ramus's dismemberment of the text, comparing it in his 1619 *Praelectiones* to the mob's rending of the body of the female mathematician Hypatia (coincidentally the daughter of Theon of Alexandria). According to Savile, the *Elements*, like Hypatia herself, was a most "beautiful body," unified and whole, though blemished by "two moles:" the theory of proportion and the parallel postulate.

Savile's statement, taken out of its polemical, even sexualized context, would become his single best-remembered contribution to mathematics, inspiring Girolamo Saccheri in his 1733 *Euclides ab omni naevo vindicatus* to attempt to rid Euclid "of every mole" – thereby quite unknowingly laying the ground for non-Euclidean geometry.

## Drowning by Numbers

Aristotle tells us that mathematics began in ancient Egypt. After the annual floods of the Nile, the learned priests restored drowned boundaries and landmarks, using the properties of triangles and squares. Josephus, on the other hand, imagined that mathematics itself survived a flood, *the* Flood, its principles having been inscribed by the sons of Seth on stone and brick pillars before God drowned most of the human race. There is a nice symmetry between the two accounts, pagan and Jewish. In the one story, mathematics saves civilization from drowning; the pure, eternal patterns of geometry can undo the disorder of the world. In the other, we are reminded that mathematics, even if eternal, depends for its existence on mere flesh which is, unfortunately, even more prone to flux and decay than boundary stones.

This is a lesson that Hippasus the Pythagorean would have done well to have learnt. He made public (some stories said) the very thing that his brethren wanted above all to keep secret: that the square root of two was irrational, so that not everything in the world could be built of whole numbers.[13] His fellow philosophers drowned him, but in vain, of course: mathematicians may drown easily, but, so long as *some* remain dry, mathematics is hard to extinguish.

To the ancients, mathematics seemed peculiarly vulnerable to the dangers of the deep. According to another oft-told tale, after a shipwreck the Cyrenaic philosopher Aristippus found himself washed up on an unknown shore. Although he had always mocked mathematics as a pointless and obscure subject, when he saw geometers' figures drawn in the sand, he assured his companions that they had reached a civilized land. According to some versions of the story, Aristippus went on to advise Greek parents that they should give their children things that they, too, could carry from a shipwreck, by which he meant mathematics and the other liberal arts.

Most of these slightly melancholy stories figured prominently in humanist accounts of the history of mathematics. They seem particularly appropriate to the humanist project. All the authors who will appear in this book were engaged in preserving and restoring the broken remnants of antiquity. Francis Bacon may have complained that the shipwreck of the ancient world left only the lightest and most trivial things floating on the surface of the water.[14] These stories seemed to promise that mathematics, an altogether weightier subject, might yet be raised from the

---

[13] Euclid (1926, vol. 1, p. 411).

[14] *New Organon* I.77.

watery grave of the ancient world. Perhaps it was enough to have the fundamentals of the art, in works such as Euclid's *Elements*; the rest could be recovered by human effort, just as mathematics had been rebuilt once before after the pillars of Seth were rediscovered. Or perhaps mathematics itself had a tenacity that could survive any shipwreck: it was a possession that any human mind could carry out of disaster, one that might even (like the geometry of the Egyptian priests) undo the effects of the calamity of cultural loss.

# Chapter 1
# Lineages of Learning

## Introduction

In his account of the Egyptian king Sesostris, Herodotus recorded that the annual flooding of the Nile often left farmers owning less land and, therefore, owing less tax. To adjudicate disputes, Sesostris commissioned a team of inspectors to judge the amount of land that had been lost, and so to assess the new tax on the land. "I think," added Herodotus, "this was the way in which geometry was invented, and passed afterwards into Greece – for knowledge of the sundial and the gnomon and the twelve divisions of the day came into Greece from Babylon."[1]

Even Herodotus here admitted that he was speculating on the origins of geometry. This was, at best, a likely story to explain the rapid development of geometry in Herodotus's own time – perhaps founded on some genuine awareness of the Egyptian origins of mathematics, but equally reflecting the Greek fascination with all things Egyptian.[2] Whatever its source, Herodotus's conjecture would be repeated with ever greater certainty by later writers. The historian's veracity was strengthened by a remark by Aristotle at the beginning of the *Metaphysics*. There he explained that arts that went beyond utility or pleasure – the speculative sciences – were discovered when human beings had the leisure to go beyond their immediate, bodily needs. So it was, he wrote, that "the mathematical arts were founded in Egypt; for there the priestly caste was allowed to be at leisure."[3]

These two superficially similar stories are, in reality, quite contradictory. According to Herodotus, geometry arose out of immediate needs, and was put to work to solve unglamorous problems in land taxation. For Aristotle, in contrast, mathematics came about as a kind of mental diversion, once the necessities of life had been taken care of. As I shall argue throughout this book, these were the two responses to mathematics and its origins that would continue to characterize the scholarly opinion, including that of sixteenth-century humanist writers on mathematics (who, nevertheless, would often seek to collapse these two stories together). In this

[1] Herodotus, *Histories* 2.109.

[2] See Momigliano (1975).

[3] *Metaphysics* I.1 (981b22–24).

R. Goulding, *Defending Hypatia*, Archimedes 25,
DOI 10.1007/978-90-481-3542-4_1, 

chapter, I will survey the ancient accounts of the history of mathematics that would be crucial to these scholars' work, and also look at several important early Renaissance responses to these texts, in which were established some "standard narratives" that were all but taken for granted by later generations of mathematical humanists.

## Diodorus Siculus

An important source for the early history of mathematics was the universal history of Diodorus Siculus, written in the first century B.C. Diodorus, like so many other Greeks, was fascinated by the mythical antiquity of Egypt; and (like Herodotus and Aristotle), he attributed the invention of mathematics to Egyptian priests, who had used their leisure time to cultivate geometry, thereby solving the recurrent problem of land boundaries washed away by the flooding of the Nile.[4] Diodorus also attributed the invention of the various mathematical sciences to Hermes,[5] Sasychis[6] and the residents of Thebes,[7] without attempting to reconcile these traditions. After all, absolute precision was not his goal. Rather, Diodorus left the Greek reader with an impression of the almost unimaginable antiquity of the sciences as cultivated by the "barbarian" peoples. Just as Plato had said in the *Timaeus*, Diodorus saw the Greeks as mere children: the Chaldeans, to give the most striking example, had astronomical records going back 473,000 years.[8] (Renaissance authors would cite this number with puzzlement and skepticism, since it exceeded the age of the world according to the biblical account.) In addition, Diodorus emphasized the extraordinary *accomplishment* of these ancient barbarians: they surpassed all their successors – including the Greeks – in their knowledge of astronomy, mathematics and all the other kindred arts.[9]

In Diodorus's account, the Egyptians formed the root of an ethnic tree whose branches extended throughout the civilized world. The Jews and the Babylonians were sent out from Egypt as colonists. The Chaldaeans, the most astrologically accomplished of all nations, were simply Egyptian priests transplanted to a new land.[10] In this way, the superlative scientific accomplishments of the Egyptians were transmitted largely intact from people to people. This *translatio studii* eventually, in comparatively recent times, reached as far as the Greeks. Diodorus recorded visits to Egypt by Orpheus, Homer, Pythagoras, Solon,[11] Democritus,[12] Lycurgus and

[4] Diodorus Siculus, *Bibliotheca historica* I.81.2.

[5] *Ibid.*, I.16.1.

[6] *Ibid.*, I.94.3.

[7] *Ibid.*, I.50.1.

[8] *Ibid.* II.31.9.

[9] *Ibid.*, I.81.4.

[10] *Ibid.*, I.28.1–2.

[11] *Ibid.*, I.69.4.

[12] *Ibid.*, I.98.3

Plato,[13] each of whom brought back elements of Egyptian culture and learning for the fledgling Greek nation. Diodorus attributed even the discovery of basic astronomical facts – for instance, that the sun moves on an oblique course, in a direction opposite to the daily motion of the stars – not to Greek ingenuity but to barbarian wisdom transported to Greece.[14]

Diodorus wrote for a Greek audience within the Roman Empire, curious about the many peoples that made up their new *oikoumenê*. The lure of Egypt was, if anything, stronger than in Plato's day, and Diodorus aimed to satisfy his readers' curiosity about this most exotic of ancient civilizations. In doing so, he may seem to have slighted the native achievements of the Greeks. It is certainly true that Diodorus paid little attention to the development of the sciences in the Greek world; nor did he consider the observational astronomy of the Egyptians or the Babylonians to be in any way different from the theoretical conclusions of, say, an Aristarchus or Hipparchus. But his point was to emphasize the *antiquity* of all this learning. In Diodorus's account, the Greeks had inherited their learning from others; when they practiced mathematics, they took their place in the great genealogy of knowledge, stretching back over thousands of years. For Renaissance humanists, themselves the beneficiaries of much older civilizations, Diodorus's narrative was compelling.

## Josephus

Another source that Renaissance humanists cited in almost every history of the sciences was Josephus's account of Jewish history (and prehistory) in his *Jewish Antiquities*. Writing a century or so after Diodorus, Josephus shared with the earlier historian an interest in the origins of the sciences – and, like Diodorus, he too insisted on their remarkable antiquity. Both sources would thus be of great interest to later apologists and historians of the sciences. Nevertheless, the two sources contradicted each other in many places, especially (as we might expect) over the role of the Jews in the transmission of knowledge.

For Diodorus, the Jews were merely one of several colonies sent out by the Egyptians, overshadowed entirely by both the metropolis (Egypt) and their fellow colonists, the Babylonians and Chaldeans. Josephus all but inverted Diodorus's narrative: in his account the Jews anchored the chain of transmission, passing on their knowledge to the rest of the world, *including* the Egyptians. Josephus' bold transformation of his people – from mere offshoots of a great empire to teachers and benefactors of all nations – contributed to his larger apologetic mission. He had set out, explicitly in his *Contra Apionem* and implicitly in the *Jewish Antiquities*, to answer Greek criticisms that the Jews had contributed no arts or sciences to

[13] *Ibid.*, I.98.1.

[14] *Ibid.*, I.98.3.

the world.[15] He answered this cultural criticism through a creative re-reading of the historical texts of his people, in particular the histories and genealogies of the Hebrew Scriptures.

According to Josephus's version of patriarchal history, Seth, the son of Adam, was the very first scientist. With his sons, he conducted systematic observations of the stars, thereby establishing the science of astronomy.[16] Josephus even suggested that God had given Seth and the other patriarchs such prolonged lives so that they could make the many observations needed to ground the science of the stars.[17] In the midst of their investigation of the heavenly bodies, Seth and his family recalled that Adam "foretold" that the world would be destroyed twice, once by flood, once by fire – a statement that might suggest (and did indeed suggest to later readers) that Adam had predicted these events from the positions of the stars and was thus himself a proficient astronomer and astrologer. In order to preserve the sciences from the impending disasters, the children of Seth set up two pillars, on each of which they inscribed a complete account of the sciences. One was made of stone, to withstand the flood, the other of brick to survive the fiery cataclysm. The stone pillar, Josephus assured his Greek readers, did indeed survive the flood and could still be seen "in the land of Seiris."[18]

Josephus did not mention the pillars again, but it can be presumed that once the flood-waters receded, Noah found the stone pillar and thus rediscovered the astronomy of the ancients. By the time of the patriarch Abraham – a crucial figure in Josephus's narrative and in most accounts of the history of the sciences that followed him – the science of the stars had spread through the world again. According to Josephus, Abraham was a Chaldaean, a people fabled for their knowledge of astronomy and astrology and notorious for their worship of the stars. He was steeped in the astronomical learning of his people, but was led by it to a conclusion quite contrary to their religious beliefs. Considering the tortuous motions of the planets through the sky, Abraham reasoned that *if* they had been intelligent, they would have chosen to pursue a more regular course. But unintelligent creatures could hardly pursue such complex paths – thus, he concluded, they must move at the command of a single, sovereign God.[19]

Josephus's story is remarkable for several reasons. The book of Genesis says nothing of Abraham's reasons for believing in God; it relates only that God spoke to him, and ordered him to leave his house and travel to a new country.[20] In Josephus's version, Abraham became convinced that God was one and separate from his creatures by the natural action of reason (and through an argument from

[15] On Josephus's apologetic program and his elevation of Abraham as a philosopher and founder of sciences (discussed below), see Feldman (1968).

[16] Josephus, *Jewish Antiquities* I.69–72.

[17] *Ibid.*, I.105–108.

[18] *Ibid.*, I.70. On Josephus's account of the earliest history of the sciences, and the sources and *fortuna* of his story of the two pillars see Lutz (1956).

[19] Josephus, *Jewish Antiquities* I.154–157.

[20] Genesis 12:1.

the *irregularity* of the universe quite unattested in any other ancient source).[21] Josephus's intent, as it was throughout the *Antiquities*, was to convince the pagan reader of the reasonableness of Jewish beliefs and thereby to win assent and even conversion to the worship of the Hebrew God. Abraham, as a man who *chose* to obey God, served Josephus's purpose very well: the polytheistic star-worshipper whom reason persuaded to abandon his pagan ways was not too different from Josephus's intended converts.

Abraham's apostasy from his ancestral faith was not popular among his countrymen, and he was driven by them from Chaldaea to Canaan (an enforced flight which, Josephus added, coincided with God's will for him). From there he moved to Judaea and then, most opportunely for the subsequent development of the sciences, he made his way to Egypt. Again, while Scripture records the barest of details – that Abraham moved to Egypt because of famine in his own land[22] – Josephus expanded on the story most significantly. It was not only famine that drove Abraham to Egypt, he wrote, but also curiosity about the religious beliefs of the Egyptians. Abraham, suggested Josephus, wanted to put his new-found monotheism to the test by hearing what the Egyptian priests had to say about their own gods: "if he found their doctrine more excellent than his own, [he intended] to conform to it, or else to convert them to a better mind should his own beliefs prove superior."[23] But he found in Egypt a chaos of competing practices and customs, all at war with one another. With an obvious nod towards Socrates, Josephus had Abraham debate with each sect until its adherents were compelled to admit that their beliefs had no solid basis. Rendered docile by Abraham's dialectic, the Egyptians looked to him for instruction in the truth; and it is then that

> he introduced them to arithmetic and transmitted to them the laws of astronomy. For before the coming of Abraham, the Egyptians were ignorant of these sciences, which thus travelled from the Chaldeans into Egypt, whence they passed to the Greeks.[24]

This was a crucial move in Josephus's apology for Jewish culture. The Egyptians, long held up by the Greeks as paragons of learning and antiquity, were in fact students of the Jews. Moreover, through his construction of a *commigratio scientiae*, the ancient learning of the Chaldeans (presumably derived from the antediluvian science of Seth and his sons, though Josephus never makes this explicit) passed through the founder of the Jewish people to the Egyptians *and then to the Greeks*. In a stroke, Josephus made the entire Mediterranean world dependent on the Jews for their learning and civilization. This narrative was destined to be repeated endlessly through the Middle Ages and Renaissance. It functioned much as the legends in the history of philosophy of a *prisca sapientia* delivered by Moses to the

---

[21] See Feldman (1968, pp. 145–149). As Feldman shows, some rabbinic sources had also claimed that Abraham discovered monotheism through rational argument; none, however, attributed this unusual argument to him.

[22] *Genesis* 12:10.

[23] Josephus, *Jewish Antiquities* I.161.

[24] *Ibid.*, I.167–168.

Egyptians and thence, via Hermes Trismegistus and others, to Plato: both traditions assured Christian readers of ancient pagan learning that this knowledge was safe, endowed with a legitimate pedigree, and entirely compatible with scripture.

## Proclus

The histories of Diodorus and Josephus treated the origin of the sciences in passing, along with many other matters. Renaissance readers also possessed another text on the history of mathematics of a quite different cast, written from a mathematically sophisticated point of view. In his *Commentary on the First Book of Euclid's* Elements, the fifth-century Neoplatonist Proclus sketched out a very brief history of mathematics culminating in Euclid and the writing of the *Elements*. The context for Proclus's digression into history was his survey, in the preface to his *Commentary*, of the wide application of geometry. This science, Proclus said, spanned the whole ontological spectrum, from the most sublime and detached speculation on eternal things (which, to Proclus the Platonist, were naturally the best objects) to applications useful to human life – here Proclus cited the well-worn stories of Archimedes launching a ship and weighing Hiero's crown as examples to show how mathematics could bring about practical benefits. Yet these impressive feats did not really hold his interest; such marvels revealed nothing about the development of mathematics. For, in Proclus's scheme, mathematics had constantly developed *from* such concrete and practical achievements to a more abstract and contemplative mode.[25]

Proclus placed the roots of mathematics quite conventionally with the Egyptians who (according to the familiar, Aristotelian story) invented geometry to restore boundary lines after the flooding of the Nile. Geometry was thus discovered out of "necessity," like the other sciences; but, also like the other sciences and everything else in this world of becoming, it subsequently grew out of imperfection towards perfection. "Thus they [sc. geometers] would naturally pass from sense-perception to calculation and from calculation to reason."[26] In the same way, Phoenicians had invented numbers to assist them with trade, and the invention then allowed for the development of arithmetic, the "accurate study" of numbers for their own sake.

Moving beyond these legendary beginnings, Proclus turned to the history of mathematics among the Greeks. For this he drew on an excellent primary source: a now lost history of geometry written by Eudemus, a student of Aristotle's. (Because of its witness to this valuable source, Proclus's historical excursus is often referred to as the "Eudemean summary").[27] Greek mathematics began with Thales, who travelled to Egypt and presumably brought geometry back with him. He tackled some

[25] Proclus (1992, pp. 50–52).

[26] *Ibid.*, p. 52.

[27] Proclus *may* have accessed the source second-hand, via a summary of its contents in a later history of geometry. See Euclid (1926, vol. 1, pp. 35–38).

problems "empirically" (as the Egyptians had done), but also began to find general solutions. Next came Pythagoras, who transformed mathematics into a liberal discipline. He surveyed the principles of mathematics from the highest downwards, and investigated theorems in an immaterial and intellectual manner. Among his discoveries were proportionals and the five regular solids.

So the list proceeds, with each new name furthering the progress of geometry from practical instrument to abstract, demonstrated science. But the crucial step came with Plato, "who greatly advanced mathematics in general and geometry in particular."[28] The subsequent mathematicians in Proclus's list are characterized by their relationship to Plato: students of his (or students of students, and so forth), or members of the Academy, or simply Platonists. Proclus probably found a preponderance of mathematicians of the Academy in his source, Eudemus, who wrote within living memory of Plato himself. But it is also clear that Proclus constructed his list quite intentionally, both to demonstrate his thesis that mathematics developed by striving for abstraction, and to connect its development with his own philosophical school. As modern historians have noted, throughout his brief history Proclus linked mathematicians together into a continuous succession of teachers and students, in precisely the same way that philosophical schools (including Platonism) claimed an unbroken lineage stretching all the way back to their founders.[29] More importantly, he imposed a metaphysical narrative onto his historical framework, in which mathematics itself progressed from the more material (the measurement of land) to the more abstract (theorems and proofs), heading ever closer to its final goal of the contemplation of intelligible realities and divinity itself. Thus, while Proclus acknowledged the barbarian origins of the sciences, he reserved their full development as abstract, demonstrated systems of knowledge for the Greeks, and attributed their complete fruition to mathematicians associated with the Platonic school. Even Euclid (about whom, as we shall see, Proclus had almost no certain information)[30] was enlisted as a member of the Platonic school.

Proclus's *Commentary* was first published in Greek in 1533, alongside the *editio princeps* of the Greek *Elements*; in 1560, it appeared in an excellent Latin translation.[31] Abbreviated though it is, the importance of Proclus's historical account of geometry (the only narrative history of Greek mathematics to survive from the Greek world) cannot be overstated; from the moment of its publication it exerted an influence on the historiography of mathematics quite out of proportion to its length. In this respect, Proclus's *Commentary* differed from the writings of Josephus and Diodorus, whose ideas were already widely spread through Eusebius's *Praeparatio evangelii*, medieval chronicles and other sources long before humanists turned their attention to the original texts. In contrast, there was a definite change in the content and character brought about by the new presence of Proclus (traits that can already

---

[28] Proclus (1992, p. 54).

[29] Cuomo (2001, p. 56).

[30] See fifth chapter, at note 20.

[31] Euclid (1533); Proclus (1560).

be seen in the handful of humanists who knew the work in manuscript before it was printed). The bare historical data – the list of names of Greek mathematicians and some basic details of their discoveries – would be repeated endlessly, compared with other sources and expanded upon, sometimes quite imaginatively. But beyond this, Proclus also affected the *tenor* of the history of mathematics. Both his association of mathematics with philosophy, and his equation of abstraction with progress would be taken for granted by many Renaissance writers on the subject. The history of mathematics took on a Platonic cast.

There were many other sources that humanist historians drew upon for their accounts of the development of mathematics, of course. Authors as diverse as Pliny, Diogenes Laërtius and Cicero preserved anecdotes about ancient mathematicians that Renaissance historians would mine for their accounts. These sources were often the fruit of the individual historian's research, appearing in some histories only to be ignored in others; they will be examined in detail in the chapters below, introduced at the points where the humanist historians themselves found them useful.

In contrast, Diodorus Siculus, Josephus and Proclus formed the basic core of most Renaissance narratives. They were important in part *because* they conflicted with each other in fundamental ways. Diodorus elevated Egyptian priests and the Chaldeans as the first mathematicians, of quite unimaginable antiquity. Josephus insisted that the Jews had first developed the sciences, and were the teachers of both the Chaldeans and the Egyptians. Both authors agreed only that the Greeks had derived their knowledge second-hand from these barbarian peoples. Proclus, meanwhile, cared less about the primitive origins of the art and more about its natural progress from practical application to philosophical abstraction. The apparent incommensurability of these major sources motivated Renaissance historians to seek out other sources that might help to reconcile their narratives.

This book focuses on two great histories of mathematics, by Peter Ramus and Henry Savile. Yet they were hardly the first Renaissance scholars to write on this subject; while Savile and Ramus wrote on a scale and depth that surpassed their predecessors, they were certainly aware of earlier writings and often drew upon them. In the remainder of this chapter, I will discuss briefly some of the more important texts that preceded Ramus and Savile's histories. Other texts (including medieval sources) that were drawn upon for particular elements of historical narratives will be introduced in later chapters where they are relevant.

## Johannes Regiomontanus

In 1464, the great German mathematician Johannes Regiomontanus delivered a set of astronomical lectures to the faculty at the university of Padua, based on al-Farghani's elementary handbook of astronomy in its medieval Latin translation. He began his course with a historical survey, quite brief (11 pages printed, or about 4,000 words), but covering the history of the discipline from antiquity to his own day. This small tract deserves to be called the first modern history of mathematics, and has been the subject of recent scholarly study.[32] Regiomontanus was a skilled reader and editor of ancient texts, in many ways the archetypal "mathematical humanist." The oration, too, is a fine piece of classicizing Latin and humanist rhetoric. Yet, as James Byrne has amply shown, this work does not fit into the humanist genre quite as readily as one would suppose.[33] In particular, Regiomontanus took little interest in the sources I have described in this chapter, neither in their tales of fabulous antiquity nor in Proclus's lineages of mostly forgotten mathematicians.

Regiomontanus, a highly original and creative mathematician who had dedicated himself to a program of translation and publication, intended to convince his audience not just of the antiquity of the sciences, but also of their current vitality. He began his history promising not just to reveal the origins of the arts and their passage from nation to nation, but to tell "how they were at last translated from various foreign tongues into Latin, which of our ancestors were famed in these disciplines, and to which moderns recognition should be granted."[34] Like Aristotle and so many other ancient authors, Regiomontanus found the origins of geometry in Egypt, in the aftermath of the annual flood of the Nile. He followed Herodotus (with some notable variations), in emphasizing the service that the foundation of geometry had offered to the state. After the floods, he wrote, squabbles would break out among the farmers, as they tried to enlarge their holdings by arguments and threats. Finally, the king set down some methods whereby they could come to amicable agreements. Thus, under political pressure to settle their differences, men were pushed "by a widespread yet uncommon impulse" (*generali et inusitato quodam impulsu*) to think about measurements, and then to put their discoveries into some sort of order and commit them to writing. Texts of this genre made it into the hands of "Euclid of Megara,"[35] who added some innovations of his own to produce the *Elements*.[36]

Regiomontanus's narrative is remarkable for the continuity it asserts between the geometry of Euclid and the practical science of the supposed inventors of geometry. The art passed directly (or, at least, via a fairly primitive set of texts) from the Egyptian surveyors to Euclid. There is no sign of Pythagoras who so often appeared as an intermediary bringing back the wisdom of the Egyptians to the Greeks. Regiomontanus *did* mention Pythagoras later in the oration, when discussing the origins of arithmetic, but only to dismiss him in a single clause:

> For even if Pythagoras's skill in numbers was legendary to all who came after him, both because he made himself the student of foreign teachers, Egyptians and Arabs, who helped

---

[32] See Byrne (2006). Swerdlow (1993a) is a paraphrase of the oration with commentary and extensive bibliography. The edition referred to is Alfraganus (1537); there is a facsimile reproduction of Regiomontanus's speech from this edition in Regiomontanus (1972, pp. 43–53).

[33] Byrne (2006, especially pp. 51–56).

[34] Alfraganus (1537, sig. $\alpha$4r): "quo pacto ex linguis peregrinis variis ad Latinos tandem pervenerint, qui in hisce disciplinis apud maiores nostros claruerunt, et quibus nostra tempestate mortalibus palma tribuitur."

[35] The reason for this epithet will be explored in fifth chapter.

[36] *Ibid.*, $\alpha$4v.

> him a great deal, and also because he tried to hunt out the secrets of nature, relying on the solid basis of numbers, nevertheless Euclid laid a much worthier foundation for numbers in three books, the seventh, eighth and ninth.[37]

The utility and mathematical excellence of Euclid's *extant* books far outweighed the mystique of Pythagoras, who merely transmitted the accomplishments of the Egyptians and was a liminal figure, himself almost as legendary as the mathematicians of the deep past. Regiomontanus was also aware of Josephus's narrative, but reported it without much confidence: "They also proclaim that Abraham, father of the Hebrews, had some astronomy; Moses too."[38] In Regiomontanus's view, Greek and Latin ingenuity had far surpassed prehistoric wisdom. It was Euclid and Jordanus de Nemore – a medieval arithmetician! – who were the *true* founders of the arts, their precise and ordered collections of theorems preferable to the mystical and mythical achievements of Pythagoras. Similarly, the Hellenistic astronomers Hipparchus and Ptolemy deserved the real credit for inventing astronomy, not Abraham, or Prometheus for that matter. After all, Hipparchus discovered the precession of the equinoxes, without which there could scarcely be an adequate astronomy or calendar.[39]

Taking Regiomontanus's history as a whole, the names of historical figures far outnumber the legendary progenitors of the arts. Among the historical figures, ancient mathematicians and his own contemporaries received equal treatment. Those whose books were still extant and which he had found illuminating or useful received particular attention. In short, Regiomontanus's oration focused on mathematical *practice* – and on a kind of practice that went far beyond the limits of university mathematics of the day.[40] Like the later authors whose histories will be explored in depth in later chapters, Regiomontanus fashioned his history of mathematics to suit his vision of mathematics as he practiced it himself – and as he hoped others would practice it.

## Polydore Vergil

In 1499, the humanist Polydore Vergil briefly surveyed some accounts of the origins of the sciences in his massive encyclopedia of the origins of things, *De inventoribus rerum*.[41] His collection of opinions about the origins of astrology and geometry

---

[37] *Ibid.*, sig. $\beta$1v: "Nam etsi Pythagorae numerorum peritia apud posteros immortalitatem reliquerit, tum quod peregrinis praeceptoribus Aegyptiis atque Arabibus, qui plurimum in eo studio valuerunt, se submiserit, tum quod numerorum certa compagine omnia naturae secreta scrutari tentaverit, longe tamen digniora Euclidea iecit numerorum fundamenta in tribus libris suis, septimo, octavo, et nono."

[38] *Ibid.*: "Abraham enim Hebraeorum patrem Astronomiam tenuisse clamant atque Mosen."

[39] *Ibid.*

[40] Byrne (2006, pp. 56–57).

[41] See Popper (2006, pp. 92–93); and, on Vergil in general, the introduction to Vergil (2002) and references found there; all quotations here are taken from that edition.

demonstrates just how chaotic the narratives were that humanists inherited from the ancients. Opening his account of astrology,[42] Vergil cast a jaundiced eye over the claims of astrologers, such as the fourth-century apologist for the science Firmicius Maternus: that a certain horoscope would create an accountant, another a charioteer. Such extravagant claims for the stars seemed to Vergil to accord with the religion of the Egyptians, who had indeed claimed the invention of this science, according to Diodorus Siculus. Yet the same author, Vergil observed, also attributed the origin of astrology to the Chaldeans, and to Mercury, and to Actinus the son of the sun. Josephus, on the other hand, maintained that Abraham, a Chaldean, first taught astrology to the Egyptians, and that Greek philosophers such as Thales and Pythagoras, who introduced astronomy to their own people, received their learning from the Egyptians and Chaldeans. Turning to Pliny, Vergil found mathematics attributed to Atlas, Jupiter Belus,[43] the Phoenician people and the Assyrians. On the other hand, Servius, in his commentary on Virgil's *Eclogues*, recorded that the Assyrians received their knowledge from Prometheus.

Vergil harmonized these accounts by assuming that there must have been several later discoveries, or rediscoveries of astrology. The true, and original discovery was that of the sons of Seth, as related by Josephus. This was unquestionably the earliest discovery, for it came shortly after the creation of the world, and it was recorded by "Josephus, a most important authority indeed" (*teste Iosepho autore sane gravissimo*). So, by inference, it was worthy of credence. Vergil went on to retell the story of Adam predicting the future destruction of the world by flood and fire, and the construction of the two pillars to preserve the knowledge of the stars for later generations. Having accepted Josephus's account, Vergil constructed a tiny *commigratio* of knowledge, whereby the ancient knowledge reached later peoples:

> And so it is reasonable to believe that astrology came from the Hebrews to the Egyptians and Chaldaeans, and from them to other peoples. Such was the beginning of the art of astrology, which doubtless was devised simply to befuddle sound minds.[44]

Despite finding the origin of astrology in the very dawn of the human race, Vergil does not seem to have warmed to it at all. Nevertheless, he went on to list a few miscellaneous astronomical "firsts:" the first explanation of an eclipse (Anaxagoras), the first person to realize the morning and evening stars were the same (Parmenides), and the first mechanical model of the heavens (Archimedes, Musaeus, Anaximander or Atlas, according to different authorities). His sources for this information were thoroughly humanist texts: Pliny, Cicero and Plutarch's *Lives*, which he used indiscriminately (here and elswhere in his work) as sources of information, with no real attempt to assess the relative trustworthiness or seriousness of their accounts; they were treated merely as funds of anecdotal data.

---

[42] Vergil (2002, I.17).

[43] That is, the Babylonian Bel or Canaanite Ba'al; there may be more uniformity in Pliny's statements than Vergil has perceived.

[44] Vergil (2002, I.17.5).

In his next chapter, Vergil turned to the invention of geometry and arithmetic.[45] He retold the story of the Egyptian invention of geometry as a way to restore boundaries after the flooding of the Nile – a story he expanded with interesting but irrelevant information from ancient geographers about this annual phenomenon. Most ancient authorities claimed that the Egyptians invented geometry, and that the Phoenicians (driven by commercial, not geographical necessity) invented arithmetic. But Vergil again preferred the account of Josephus, who attributed both sciences to the ancient Hebrews. Vergil recalled Josephus's account of the long lives of the patriarchs, which they devoted to the discovery of astrology and geometry, and Abraham's role in transmitting this knowledge to the Egyptians. The Egyptians had been entirely ignorant of geometry until Abraham arrived in their country – a situation that Vergil did not try to reconcile with the story that the Egyptians had long relied on geometry to restore their boundaries after the annual flood.

Boundary marks also come up in the next chapter, on weights, measures and numbers.[46] As before, Vergil catalogued several mutually inconsistent accounts of the origins of measurement, before once again deciding that these all represented later, perhaps independent rediscoveries. For the earliest origins, he again turned to Josephus who recorded that Cain, the wicked son of Adam, had first divided up the earth with boundary marks – this account is what one ought to believe (*ut credere convenit*).

In his account of the origins of the sciences, Vergil was more concerned to parade his erudition than to arrive at a coherent historical narrative. In this respect, he was typical of many humanist writers of his and later generations. To the extent that he had a guiding principle in his historiography, it was the pious exaltation of the Hebrews over all other nations. Beyond simply marvelling at the great antiquity of the arts, Vergil intended to secure a place for the ancient speculative arts in contemporary Christian society by providing them with a quasi-biblical pedigree.

## Girolamo Cardano

In sharp contrast to Polydore Vergil, the Italian mathematician Girolamo Cardano took a quite dim view of the Josephan histories.[47] To his work on the arithmetic of whole numbers, Cardano prefaced a brief history of mathematics in which he wrote that the origins of arithmetic had been lost in the depths of time, with the result that each nation laid claim to the honor for themselves. Such patriotic pride had led the Jews, too, to claim credit for the invention. Quite uniquely among historians,

[45] Vergil (2002, I.18).

[46] Vergil (2002, I.19).

[47] On Cardano's other works of mathematical history, not treated here, see Grafton (1997, pp. 270–271).

Cardano connected Josephus with stereotypes of contemporary Jews, giving his skepticism, or even irritation with such stories an unpleasant anti-Semitic edge:

> The Jews, in their always hollow zeal, brag about their superstitious reverence for antiquity (they have, after all, almost nothing else to be proud about). They say that the art of calculating was invented by those grandchildren of Adam who came from the line of Seth...[48]

Cardano allowed that there *might* be some truth in this story, as there might be in the accounts of Abraham or Mercury as founders of mathematics; but the first fact on which one could depend (*illud satis constat*) was that Pythagoras had brought arithmetic to the Greek world, where Euclid then developed it to a high level of sophistication. Cardano peppered his account of the subsequent history of the science with references to manuscripts he had discovered or books by great practitioners of the art in his own collection. In short, Cardano focused on authors and texts that could be *used* in the practice and development of arithmetic. He devoted the main portion of his history to medieval Latin and Arab and modern arithmeticians, whose writings were still extant.[49] With such a practical focus, it is not surprising that Cardano was dismissive of stories about the antediluvian origins of mathematics. In this respect, Cardano resembled Regiomontanus, whose oration he likely knew.

Cardano engaged with the history of mathematics again in an "encomium of geometry" which he delivered in 1535 in Milan.[50] A major theme of his panegyric was the universality of geometry. It had such a vast reach that it deserved its name of "earth measurement," its range of application as wide as the earth itself.[51] This conceit led him to consider the actual origin of the name "geometry" and the history of the science. However broad its modern use, geometry was invented from the need to measure land, and most likely can be attributed to "our first parents" who lived near the banks of the Nile. It is remarkable that, in his enthusiasm for the Egyptian origin of geometry and his distrust of Jewish sources, Cardano not only ignored the Josephan account but even the existence of human beings prior to Egyptian civilization. Aware that some might question his starting point, he suggested that the first human being himself might have dwelt in Egypt, or perhaps some calamity reduced the human race to only a few Egyptians.[52] In either case, the origin of mathematics was linked to the flooding of the Nile, and no place was left either for Josephus's fanciful stories or for any contribution by the Jews to European knowledge. As in his arithmetical history, Cardano had little patience for arguments

---

[48] Cardano (1663, vol. 10, p. 118): "Iudaei cultum antiquitatis superstitiosum (ut qui pene nulla alia ex parte gloriari possint) semper inani studio iactantes dicunt ab Adami nepotibus, qui ex Seth prodiere, inventam esse numerandi artem..."

[49] Rose (1975, p. 143).

[50] Cardano (1663, vol. 4, pp. 440–445).

[51] *Ibid.*, p. 442.

[52] *Ibid.*: "Seu enim unus atque primus homo fuerit, seu per aliquam calamitatem in eam paucitatem ventum sit, constat illos primos parentes nostros vicina Nilo Aegypti loca incoluisse."

*de originibus*. "It is useless for anyone to ask me when [geometry] was invented, or by whom,"[53] he exclaimed. But the obscurity of its origin was no obstacle to its dignity, he insisted, alluding to and rejecting the traditional humanist veneration for ancient pedigrees. Nevertheless, Cardano went on to note that some had claimed that geometry was founded at Creation – or even that it was eternal and governed creation itself. Thus, concluded Cardano, antiquity was ennobled by the presence of geometry, while geometry itself needed nothing from antiquity.

In his *Encomium*, Cardano drew information from Diodorus Siculus, whose fabulously ancient history of Egypt he reported with much skepticism. In addition, Cardano knew Proclus's *Commentary*, published only 2 years previously, which he used to fill out the history of early Greek geometry.[54] Cardano clearly also shared Proclus's lofty conception of the nature of mathematics. After a survey of ancient Greek, medieval and modern mathematicians, Cardano returned to the creation of the world. Here he offered a summary of Plato's *Timaeus*, emphasizing the use of geometry in each stage of the making of the universe. It is because geometry has such a sublime aspect, he concluded, that "no philosopher, no prince has wanted to be without this science."[55] Ultimately Cardano's suspicion of the grandiose claims made for the prehistory of mathematics led him to an exalted Platonism that transcended history altogether.

## Melanchthon

The lure of Platonism was always strong, even for an Aristotelian like Philip Melanchthon. In a widely-read encomium of mathematics first published in 1536, the reformer began and ended his praises of the art with the famous sign over the door of Plato's Academy: "let no one enter who is untrained in geometry."[56] Melanchthon suggested two interpretations of Plato's sign. First, the philosopher may have intended mathematics to be a prerequisite for the study of the other arts, particularly philosophy. For although geometry had its uses for practical men who made buildings or pots, ultimately it was the philosopher who really needed it. In a passage that would resonate with later readers, Melanchthon claimed that the natural

---

53 *Ibid.*: "frustra quis a me requirat, quando inventa sit, vel a quibus."

54 See p. 128, on Cardano's reading of Proclus in order to fix (erroneously) the dates of Euclid.

55 *Ibid.*, p. 445: "nullum philosophum, nullum principem hac scientia carere voluisse."

56 Melanchthon's praise of geometry was prefaced to Vögelin (1536), an elementary work of geometry which went through many editions in the early sixteenth century. It enjoyed even wider circulation after it was reprinted virtually unchanged the following year as the preface to an important edition of Euclid's text (Euclid 1537, reprinted in 1546). At the same time, it was printed in a collection of Melanchthon's mathematical prefaces (Melanchthon 1537). The Latin text of the Vögelin preface is in Melanchthon (1834–1860, vol. 3, cols 107–114) (to which all further references will be made). Translations of the Euclid preface are in Moore (1959) and Melanchthon (1999, pp. 98–104). On Melanchthon's mathematical prefaces in general see Methuen (1996, especially p. 388).

philosophy and physics of Aristotle could not be understood without a grounding in mathematics.[57] Moving beyond the natural world, Melanchthon noted that mathematics drew men out of their concern for worldly things to a consideration of the heavens: at first, quite literally through the measurement of the stars, but eventually "it [carries] the aspiring souls back to their homeland and into consort with celestial beings – even to the vision of God."[58]

It need hardly be emphasized just how Platonic both these arguments are. In particular, the claim that the natural philosopher needed to be trained in mathematics was quite contrary to the Aristotelian conception of physics. Melanchthon's view that mathematics was a vehicle for the elevation of the soul was also thoroughly Platonic. Melanchthon went on to consider a second interpretation of Plato's sign which, again, emphasized the effect of the art on the practitioner rather than its utility in the world. Ethical behavior was, as Plato showed, a kind of geometrical harmony;[59] thus mathematics and a knowledge of geometrical harmony in itself was necessary for the achievement of virtue. Underscoring this lofty, intellectual understanding of mathematics, Melanchthon went on to say that he was not addressing those who have no care for the liberal arts, nor those who aimed only for mercantile gain. Both types of men, warned Melanchthon, were entirely *ageômetrêtoi* and prohibited by Plato from entering the academy. He hoped that anyone who opened the book in which his preface appeared would be reminded by his opening words – "let no one enter who is untrained in geometry" – that they should always aim for the most sublime uses of this art, and not linger in calculation and measurement (even though these, too, were salutary for the well-formed mind).[60]

It was in this context that Melanchthon then turned to historical anecdote, and related the story of the wreck of Aristippus, the follower of Socrates and founder of the Cyrenaic school:

> When Aristippus lost everything he owned in a shipwreck, but nevertheless reached the shore of Rhodes in safety along with a few companions, the story goes that, while walking along the beach, he noticed some geometrical figures in elaborate constructions. Although the sea had stripped them of all their provisions and thrown them up onto some unknown land, once Aristippus saw these figures he bid his companions to be of good heart, saying that he had seen the footprints of men, and was glad for himself and the others because they had not been washed up onto some barbarous shore; and he assured them that humanity towards shipwrecked strangers would not be wanting in men who cultivated the study of these arts. How I wish that those footprints of men which Aristippus marvelled at on the

[57] Melanchthon (1834–1860, vol. 3, col. 108). Regiomontanus made a similar argument in his oration; see Byrne (2006, pp. 57–58). For the influence of this element of Melanchthon's thought on Ramus, see p. 69.

[58] *Ibid.*: "Denique exultantes animos in patriam ac familiaritatem coelestium atque adeo ad agnitionem Dei traduxit."

[59] Melanchthon (*ibid.*, col. 108) refers to Plato's *Gorgias* (508A), where Callicles is told that his undisciplined morality is due to his neglect of geometry. Later in the encomium, Melanchthon also cited from the *Republic* the geometrical harmony of the ideal state and ideal soul.

[60] *Ibid.*, col. 109.

shore were more frequent in our schools. For these arts have lain deserted and neglected for many centuries now. [61]

The story of Aristippus had originally appeared in one of the prefaces to Vitruvius's *De architectura* (Fig. 1.1).[62] In that version, Aristippus (who infamously criticized mathematics as trivial)[63] recognized in the diagrams the presence of educated men and, by inference, the prospect of lucrative employment for a philosopher. He took himself to the local gymnasium and was soon able to recover his property and more. As his companions (now reclothed by his largesse) made to leave Rhodes, Aristippus advised them "that children ought to be provided with property and resources of a kind that could swim with them even out of a shipwreck." Canny advice, no doubt, but not entirely in the spirit of Melanchthon's elevated conception of mathematics. From this striking story, Melanchthon retained the notion that mathematics was a mental possession, but otherwise quite recast the intention of the anecdote. The pursuit of geometry, he argued, revealed a people who had achieved a high level of *moral* mastery, along with intellectual culture. They would be just, and would know how to show humanity to the lost – a conceit that Melanchthon brilliantly turned around to a criticism of the state of the contemporary university.

This is the extent of historical engagement in Melanchthon's preface. Nevertheless, this frequently printed encomium deserves mention because of its influence on later writers on the history of mathematics. The strong Platonism would remain a feature of such histories (or, in the case of Ramus, something which the historian had quite explicitly to resist). His retelling of the Aristippus story would be picked up by several later writers, and would eventually become virtually symbolic of Greek mathematics. Moreover, Melanchthon's artful reshaping of a historical anecdote to suit his own purposes and, especially, to comment on the contemporary state of the discipline, established a pattern which later histories of mathematics would follow again and again.

---

[61] *Ibid.*, sig. B3r: "Aristippum ferunt, cum amissis naufragio fortunis omnibus, ipse tamen cum paucis ad littus Rhodium salvus pervenisset in tabula, ambulantem in littore, geometricas figuras in machinis quibusdam conspexisse. Quanquam autem mare et viatico eos exuerat, et in loca eiecerat ignota, tamen conspectis illis figuris geometricis iussit socios bono animo esse, inquiens se vidisse hominum vestigia, gratulatusque est sibi et reliquis, quod non in barbarum littus eiecti essent, confirmavitque humanitatem erga hospites ac naufragos non defuturam illis hominibus, apud quos harum artium studia colerentur. Vtinam vero haec hominum vestigia quae ibi in littore miratus est Aristippus, in scholis etiam frequentiora essent. Iacent enim desertae et neglecta hae artes multis iam seculis."

[62] Vitruvius, *De architectura* VI, pref. See Aristippus (1961, pp. 3–4) for other ancient retellings of this anecdote (none of which drew the moral that Melanchthon did from this anecdote).

[63] Aristotle, *Metaphysics* III.2: according to Aristippus, mathematics not only failed to produce anything of value, but unlike even carpentry or cobbling it refused to take into account the notion of value at all.

**Fig. 1.1** Aristippus on the Shore of Rhodes, from David Gregory's Euclid (Oxford, 1703)

## Conclusion

The sources on the history of mathematics that survived from antiquity were haphazard, at best. Although ancient authors had written books devoted to the history of geometry or other mathematical arts, none had survived the collapse of the ancient world. Renaissance humanists had inherited instead narratives of universal history, in which the origins and development of the arts made an incidental appearance. They also had at hand a fund of anecdotes about ancient mathematicians – and, increasingly, also their works, which sixteenth-century scholars like Francesco Maurolyco and Federico Commandino were gradually making available, and comprehensible, to a wider public.

As the very brief survey in this chapter illustrates, the texts that were at hand were frequently contradictory, or scarcely informative at all. Renaissance humanists resolved the difficulties in their sources in two principal ways: by focusing on the utility of the sciences, or by embracing a Platonic notion of the nature of mathematics – and sometimes (as in the case of Cardano) by doing both. The history of mathematics revolved around these poles – or rather, was held in tension between them. In the two principal authors considered in this study, Peter Ramus and Henry Savile, we will see two versions of the history of mathematics, each clinging to an opposite pole.

# Chapter 2
# Ramus and the History of Mathematics

## Introduction

Peter Ramus (Pierre de la Ramée, 1515–1572), was born in Picardy, the son of a once wealthy family, now severely impoverished. Despite his family's financial condition, Ramus gained entrance to the University of Paris, where he paid his way by working as a servant to wealthier students. Eventually he became a teacher, at various small colleges of the University – a life of academic obscurity that was ended by his publication in 1543 of two books based on his lectures to undergraduates, in which he subjected Aristotle to remorseless attack. This led to his being banned from teaching philosophy at the University (an interdict that was, in part, the work of his great opponent Jacques Charpentier, who will figure largely in these pages). In 1551 he was not only restored to philosophical teaching but even elevated to a regius professorship in the Collège Royal – a position he held (with some interruptions) until his murder in 1572, at the hands of rioting Parisians during the St Bartholomew's Day massacre.[1]

In this chapter, I trace the development in Ramus's writings of a historical narrative of ancient mathematics. As I will show, at the beginning of his career, Ramus thought very highly of ancient mathematics, and particularly of Euclid. He gave the art a history commensurate with this view: repeating the stories about the ancient origins of mathematics, he assumed that it had remained quite the same in its transmission from the ancient patriarchs to the mathematicians of the Greek world. Over time, however, Ramus's view of mathematics changed, becoming much more critical of the state of Greek mathematics (again, with the focus on the *Elements* of Euclid); and as his attitude towards mathematics shifted, so did his history of mathematics. He laid out this history in various prefaces he wrote to mathematical works that he published. These often sketchy, abbreviated accounts of the development of mathematics nevertheless bespeak a growing engagement with the science,

[1] On Ramus's biography see Waddington (1855) and Ong (1958a) – with caution, as each is a partisan history, in its own way. The literature on Ramus's logic and other scholarship is vast; for surveys of the field see Sharratt (1972), Sharratt (1987) and Sharratt (2000). Meerhoff (2001) has been particularly formative on my understanding of Ramus's arts. See also Goulding (2006b).

R. Goulding, *Defending Hypatia*, Archimedes 25,
DOI 10.1007/978-90-481-3542-4_2, 

in which problems in its history were becoming central to Ramus's understanding of the arts in general. In the course of these works, one can see Ramus working out his thinking on issues that would receive a full treatment in his 1567 *Prooemium mathematicum* – the full-scale history of mathematics that is the subject of the next chapter.

## Ramus and the Reform of Dialectic

Ramus was, at first sight, the least likely person to write an influential history of mathematics. He was no great mathematician himself. His sympathetic biographer Nicholas Nancel related that Ramus would spend the mornings being coached in mathematics by a team of experts he had assembled, and in the afternoon would lecture on the very same subjects.[2] Ramus was from his earliest career a logician, and remained one in all his works, whether writing on mathematics or Virgil. Moreover, he conceived of all the arts – and especially mathematics – as unchanging structures of necessarily true propositions.[3] There seemed to be little room for historical development in the sciences as he imagined them.

Yet there can be no doubt that Ramus held mathematics in particular esteem. It has been argued, in fact, that he played a crucial role in linking philosophical discourse to mathematics and in promoting the use of quadrivial reasoning in the study of the natural world.[4] The origins for his enthusiasm for mathematics are to be found in his account of the nature of the arts and critique of the curriculum of the universities; his career as a *historian* of mathematics, I will argue, was directed by problems that arose in that theory as he began to immerse himself in the sciences of the ancient world.

Ramus set out his fundamental positions in his very first printed work, the *Dialecticae institutiones* (*Education in dialectic*) of 1543, a contribution to the on-going humanist attack on scholastic logic, in which he rehearsed many of the commonplace criticisms of the university dialectic. Humanists complained that logic no longer concerned itself with real human reasoning; instead, it had become a discipline studied for its own sake, wreathed in its own incomprehensible jargon, and of no practical interest at all. The new humanist dialectics of Lorenzo Valla and Rudolph Agricola, by contrast, attempted to teach the kind of practical reasoning useful for composing a speech or letter; these scholars borrowed extensively from the rhetorical works of Cicero and Quintilian to develop a highly rhetoricized logic. Questions about the formal validity of arguments were of little interest to the

---

2 Sharratt (1975, 198–200).

3 See, for instance, Ramus (1543a, fol. 31v): "... nam si res constantes sunt et aeternae, earum disputatio, explicatioque firmis, perspicuis, necessariis argumentis addici debet." ("For if things themselves are constant and eternal, then any discussion about them, or explanation, must rely on strong, clear and necessary arguments.")

4 See Reiss (1997) and Reiss (2000).

practitioners of this new humanist logic; what mattered was whether the arguments were persuasive.[5]

In his *Dialecticae institutiones*, Ramus took the humanist reformulation of dialectic a step further. He argued that dialectic – or, in fact, any art or science – consisted of three elements: nature, doctrine and exercise. The natural workings of the mind formed the most basic and significant element, with exercise or practice coming second. The third element, doctrine, was nothing more than a record of natural reasoning; in importance it paled next to nature and practice.[6] The logic of the universities bore no resemblance to the true, natural dialectic, as Ramus argued at length in the companion volume to the *Institutiones*, the innocuously titled but exuberantly offensive *Aristotelicae animadversiones* (*Observations on Aristotle*).[7] The question, then, remained: how do we gain access to this "natural reasoning?"

Ramus's answer was surprising. He directed his reader to find a group of men – *not* scholars, but completely uneducated vineyard workers. Question them about the coming year: the fertility of the soil, the quality and quantity of the crop. "And then (he wrote) from their minds, as from a mirror, an image of nature will be reflected."[8] In the reasoned replies of these uneducated men, Ramus said, one discovered every part of logic needed for *any* purpose, whether everyday discourse or the composition of poetry: the invention of arguments, the assessment of their truth and their proper and orderly presentation. Other humanists had praised man's natural logical faculties and distrusted the artificial, but Ramus was the first to look beyond the walls of the university and the writings of the ancients to find natural dialectic at work in the world around him.

If even the uneducated possessed some grasp of the arts, then the arts taught at the university should do no more than clear away the misleading junk in the mind, and allow its natural clarity to shine through.[9] Logic should be easy to learn. For Ramus this was not merely a pedagogic ideal; a natural art, if truly natural, required only practice and minimal guidance. And if logic as it was taught was *not* easy to learn (and such were the mind-boggling complexities of scholastic logic) then that was a good sign that it was not the natural art but something "fabricated" (a *commentitium*, one of Ramus's favorite critical terms).[10]

In order to restore an art to its natural simplicity, it needed organization or "method," a term for which Ramus became and has remained famous (or notorious,

[5] See Copenhaver (1992, pp. 29–30), and especially pp. 223–225 on Valla's rhetoricizing of philosophy.

[6] Ramus (1543a, fols 5v–6r): "Comparatur igitur dialectica, sicuti vis artium reliquarum, natura, doctrina, exercitatione. ... [Doctrina] (cui perpaululum loci reliquum est) sola extrinsecus a magistris assumenda est."

[7] Ramus (1543b).

[8] Ramus (1543a, fol. 6v): "... tum ex eorum ingeniis veluti speculis imago naturae resultabi."

[9] *Ibid.*, fol. 7v: "... ut hoc artificioso quasi speculo natura formae suae dignitatem perspicere, et si qua macula sit aspersa, delere atque eluere possit."

[10] See Ong (1958a, pp. 45–47) for Ramus's use of this term, beginning with his infamous master's disputation "Quaecumque ab Aristotele dicta essent, commentitia esse."

depending on one's point of view). Although the term did not occur in the 1543 works, the same concept did, under the name "second judgment" (so called because it followed his treatment of "first judgment," or the assessment of syllogistic and other forms of argument).[11] There was, claimed Ramus, a unique way to organize any art: from the most general propositions to the most specific, an arrangement often realized by dividing and subdividing categories into two. In later writings, he would describe this process by invoking a striking image. Suppose that all of the "facts" of grammar or any other art were written out onto hundreds of slips of paper, and then shaken up together into an urn. The skilled dialectician should be able to pick out the slips one by one, and place each in its unique position in the sweep from general to particular.[12]

The conclusion must be that there is a unique, correct order to be imposed on the discovered facts of any art. Moreover, this order is *natural*, in two ways: it reveals the real structure of the world,[13] and – as Ramus argued at length in his discussions of natural capacities and art – it conforms to the structure of the human mind.[14] This was the crucial point. In the case of dialectic, the methodized art was not just a useful way to arrange the precepts of logic, but a representation of the deep structure of discourse, and hence of the human mind, the instrument of discourse and the natural source of dialectic. In exactly the same way, a methodized physics would itself be a reflection of the structure of the world. Moreover, because such a physics would have a dialectical structure, it too would conform perfectly to the human mind. And, in each case, this logical structure would reflect the order in the mind of God, who created both the human mind and the physical world that it inhabits.[15]

Ramus made explicit the metaphysical linkage of nature, God and dialectic in the central section of his 1543 *Dialecticae insitutiones*. In a dense and difficult passage on "third judgment," Ramus's contentious anti-Aristotelianism verged into an idiosyncratic form of Platonism. The third form of judgment, he wrote, was the next and final step beyond second judgment (or method). By third judgment, the entire structure of all the arts was revealed to the human mind. He explained:

> There remains the final step of dialectical judgment, concerned with perceiving that power of the human sciences which is directed towards the ultimate end of all things. Through it,

[11] Ramus (1543a, fols 27r–30v).

[12] This image first appeared in the *Dialectici commentarii tres*, issued under Omer Talon's name in 1546 while Ramus was banned from teaching philosophy. The relevant passage is translated in Ong (1958a, pp. 245–246).

[13] See, for instance, Ramus (1543a, fol. 34v). Having described the affinity of his "second judgement" with the Platonic notion of individuals emanating from ideas, he wrote: "Herein lies the most beautiful correspondence of the art with the wisdom of nature." ("Haec artis est cum naturae sapientia, pulcherrima contentio.")

[14] In the conclusion to the 1543 dialectic, he claimed that a first approximation to the untaught, natural logic could be found in his own logic, "which expresses in all its parts an image of natural dialectic – crude and unpolished, no doubt, but nonetheless a true and dependable image." (Ramus, 1543a, fol. 58r): "qui ... dialecticae naturalis imaginem quamvis rudem, impolitamque, tamen veram constantemque membris omnibus expresserit.")

[15] This point is made at length – based on a study of Ramus's *Dialectique* – in Walton (1970).

> the reward of human labor can be judged and the most excellent parent and author of all things can be recognized.[16]

The process by which one attained such an intuition was banal enough. The dialectician should begin by constructing a methodized image of all the arts together, filling in as many details as he could. Third judgment itself, the highest point of education consisted of recovering this total structure, a process which would cleanse the mind of its false beliefs and allow it to recognize its innate dialectical constitution. The newly-cleansed mind, which would naturally take up into itself the whole universe of discourse, was very nearly an image of the mind of God.[17]

Ramus illustrated his third judgment with a reference to Plato's myth of the cave. Human beings were like Plato's prisoners, compelled to look upon the shadowy play of sensory particulars. In Ramus's interpretation, the light-source behind the prisoners' heads was God; its light was human reason and dialectic. The objects casting the shadows were "the genres of things and the species contained in the arts;" and the shadows themselves, "delicate, flickering with the lightest of motions, are all the things that can be touched, heard, seen and perceived through the other senses."[18] Freeing oneself from the chains in the cave, one turned from these illusory shadows to gaze upon the "reality" of dialectic itself, spreading out like a web behind the discrete particulars of the world.

It was from this lofty summit of dialectic that Ramus invoked the highest science of all, mathematics. The mind, once freed from its shackles, would take in the arts properly and entirely for the first time. Beginning with grammar and rhetoric, it would proceed to moral philosophy and physics, and at last find rest in mathematics.[19] In the next passage, Ramus posited an interesting, two-way movement between mathematics and dialectic. The dialectician, at the height of his attainment, would approach the mathematical arts in a special way through the dialectical third judgement giving them a privileged role in understanding the world.[20] At the same time, however, the mathematical arts themselves began to take on the very qualities of dialectic itself: liberating human minds from the chains in which they were held, illuminating the world and all of the arts – in short, allowing one to transcend the

[16] Ramus (1543a, fol. 35r): "Postremus superest dialectici iudicii gradus in perspicienda scientiarum humanarum virtute ad supremum rerum omnium finem referenda positus, ut laboris humani fructus possit aestimari, et optimus rerum omnium parens, atque author agnosci."

[17] Ramus (1543a, fols 37r–v): "But when dialectic freely marvels at everything, then it will turn in towards itself and begin greatly to marvel at itself, and it will not be able to judge that it is not the divine image of the divine mind." ("Sed cum haec omnia dialectica libenter admirabitur, tum in seipsam conversa vehementius incipiet admirari, et divinae mentis imaginem non divinam non poterit arbitrari.")

[18] Ramus (1543a, fol. 36v): "Imagines medio interiectae spatio, genera rerum, speciesque disciplinis et artibus comprehensae, quarum tenues, levissimoque motu nutantes umbrae, sunt haec omnia, quae tangi, audiri, cerni, caeterisque sensibus percipi possunt."

[19] Ramus (1543a, fols 36v–37r).

[20] Ramus (1543a, fol. 39v): "Itaque cum has disciplinas lumine suo dialectica lustraverit, quanto iam plenius naturalium principia rerum, et umbrarum illarum causae cernentur."

limitations of the human condition and approach the perfect knowledge of God.[21] Mathematics was both perfected by dialectic and identical with dialectic; and the paradox which seemed to arise from this dual conception was indeed only apparent. *All* the arts were, in their deepest "methodical" structure, dialectical, as was the world which formed their subject of investigation. Mathematics, however, was the most purely dialectical, insofar as it was eternally, indisputably, and necessarily true – and everyone agreed it to be so. The fact that such a perfect science existed, and existed so *undeniably*, validated Ramus's metaphysics of art, tying together in the clearest way the action of the human mind, the world which confronted it, and the deity who was the source of both the world and the commensurate structure of the human intellect.

## The Turn to History

As I have said, such a theory of knowledge did not appear to leave much room for historical development within the arts. There was, after all, only one possible art of dialectic or mathematics, imitated from the structure of the human mind and expressed through a network of connections which obtained necessarily and eternally. Ramus's dissatisfaction with the sciences of the university curriculum, however, compelled him to face a set of historical questions. If human beings had the structure of the arts hard-wired, as it were, in their minds, one might expect that they would develop mastery of the arts spontaneously. It was a puzzle, then, that so many were unaware of this structure, that there was dissent over the nature of the arts, that (in short) people needed to rely on the arts to recover their natural skills.

In his 1543 *Remarks on Aristotle*, Ramus attempted a preliminary answer to these questions by way of a historical narrative. He wrote that the art of dialectic was first formulated (though in a crude way) by Prometheus, systematized by Zeno of Elea, and brought to perfection by Plato. These authors developed their art through observation and use; theirs was a *genuine* dialectic, just as Ramus had described it in the *Institutiones*. Yet this promising beginning was soon squandered: "Up to this point, dialectical truth, and the employment of that truth was simple and naked. Henceforth, it began to be distorted and corrupted."[22] For Ramus, the ill-employment of a natural faculty built up habits that gradually obscured the innate art. Who was to blame? The subsequent decline of the practice of dialectic (and hence the effacement, through bad education, of our natural dialectic itself) Ramus laid at the feet of Aristotle:

---

[21] Ramus (1543a, fol. 40v): "Hominem corporis exigui, velut carceris angusti custodia constrictum querimur? Mathesis liberat, seu potius hominem hac mundi universitate maiorem reddit."

[22] Ramus (1543b, fol. 2v): "Hactenus veritas dialectica, veritatisque utilitas simplex et nuda fuit, quae deinceps turbari et corrumpi coepit."

> And so, as we can know from the texts of old writers, Aristotle with his books was the first to spoil the simple truth and practice of the ancients.[23]

In a striking phrase, he accused Aristotle's craven followers of "sui desperatio:" losing faith in themselves.[24] So little did they esteem their natural abilities that they allowed themselves to become intoxicated by their teacher's claims to authority. This would become Ramus's central theme in his histories of the arts. Each art, he imagined, once possessed a primal simplicity. But corruption crept in, by the ineptitude and arrogance of a single author and his followers: they elevated personal ingenuity and fabrication over the unspoilt action of nature, eventually overshadowing it all together. It was hardly a coincidence that Ramus identified the originators of corruption with the standard university authors: Aristotle destroyed dialectic, Cicero corrupted rhetoric[25] and so forth. Students were estranged from their natures and their innate talents because the schools and universities, which should have been gently polishing the mirrors of their natures, were instead clouding them with the false so-called knowledge of the authorities.

In Ramus's account of dialectic, history was all but forced upon him. The enormous disconnect between simple, natural reasoning and the artificial problems of the modern logicians demanded an explanation, which only historical narrative could provide. At this early stage in his work, however, Ramus saw no need for a history of *mathematics* because he saw it, in its role as the ultimate end of third judgment, as essentially ahistorical. To put it another way, Ramus was delighted to find in mathematics a necessarily true and unchanging science, one which was also the purest expression of the natural dialectical order of the world (or so he thought). For Ramus, the unchanging existence of mathematics as a realized, perfect science was the clearest indication that his conception of the arts was essentially correct.

## Ramus Engages with Mathematics

Ramus's attacks on Aristotle and modern logic provoked immediate fierce opposition from within the University of Paris. In 1544, after a group of politically influential professors agitated for his removal, he was banned by royal order from teaching philosophy and both his books were suppressed.[26] It was then that Ramus first turned to a sustained consideration of mathematics; although it was the culmination of his philosophical system, it seemed he had hardly given the art much serious thought.[27] In an oration of November 1544, he announced his intention to

---

[23] Ramus (1543b, fol. 3r): "Primus igitur Aristoteles (quod ex veterum monumentis intelligi possit) simplicem antiquorum veritatem et exercitationem libris suis depravavit."

[24] *Ibid.*

[25] Ramus presented his argument against Cicero in Ramus (1549).

[26] See Waddington (1855, pp. 40–52).

[27] Ramus had most likely taught mathematics before his ban from teaching, but there is no firm evidence. A student's notes are extant, made during a course of Ramus's lectures on a textbook by Oronce Fine; but they may have been made as late as 1544. See Sharratt (1966).

devote himself to mathematics, leaving the teaching of rhetoric and grammar to his colleagues Omer Talon and Barthélemy Alexandre.[28]

In the course of the speech, Ramus expressed his admiration for the art and sketched out its history, emphasizing its great antiquity and unequalled reputation among the liberal arts. He admitted some regret at being banned from teaching philosophy; but the overarching purpose of the oration was to praise mathematics to such a degree that his turn to this art would seem an elevation rather than a retreat. In this oration, just as in his *Institutiones* of the previous year, mathematics was the most natural of arts, expressing the "free thoughts" of the mind.[29] It had, moreover, a history that surpassed that of the other arts. For, anticipating the charge that he intended to corrupt the youth by teaching them new subjects, Ramus pointed to testimonies of the ancients that demonstrated mathematics to be anything but "new." Drawing on Josephus, Ramus observed that the mathematical arts had flourished well before the Flood, while grammar, rhetoric and philosophy had all been much later inventions of the Greeks.[30] The very first human beings had cultivated these arts, which they had passed on to later civilizations:

> It was the first human beings, I say, who discovered this science. Most say that Adam, Seth and Noah discovered it; and that certain wise men among the Greeks – Pythagoras, Archytas and Timaeus – refined it. But I have labored too long over this point.[31]

The history of mathematics was not, indeed, Ramus's primary theme in this oration, as his impatience to move on to other subjects shows. But he expressed certain ideas for the first time here, which would reappear frequently in later works more focused on history. First, the idea that mathematics was natural to the human mind remained constant throughout his career. In later writings, Ramus would question whether the particular form of mathematics bequeathed to Europe by the Greeks was in fact that natural mathematics, or whether it, like dialectic, had been somehow spoiled. In the 1544 oration, however, Ramus expressed no doubts about the nature of Greek mathematics: the mathematics he intended to teach was, he said, that which both the earliest human beings *and* the wisest Greeks had practiced.

Mathematics, said Ramus, was brought to a high state of perfection by Archimedes, who devised such marvelous applications,[32] and by Plato and the mathematicians among his followers. The sign over the door of the Academy – "let no one uneducated in geometry enter here" – showed the high regard in which Plato and

[28] The text of this oration is at Ramus (1599, pp. 229–239).

[29] Ramus (1599, p. 231): "mathematicas artes invisamus, et animis (quorum sunt liberae cogitationes) velut praesentes et expositas intueamur" ("let us look upon the mathematical arts, and let us contemplate them as though they were present and put before our minds of which they are the free thoughts.")

[30] *Ibid.*, p. 235.

[31] *Ibid.*, p. 232: "Primi, inquam, homines illi propagatores generis humani, Adamus, Sethus, Noëus, invenisse; excoluisse Graecae gentis quidam viri sapientes, Pythagoras, Archytas, Timaeus existimantur; verum diutius hic immoramur."

[32] *Ibid.*, pp. 232–233.

his fellow Greeks held the art.[33] Here and elsewhere in his writings on mathematics, Ramus would single out Plato and Archimedes for the highest praise: Archimedes because of his emphasis on *use*, so crucial for Ramus, and Plato because Ramus saw him and Socrates as pioneers of the true dialectic.

Finally, there was Ramus's insistence that the mathematical arts were very old; older, in fact, than the other university arts. He would insist repeatedly in later writings that mathematics received its name (which means simply "learning" in Greek) because for many ages it was the *only* art studied. Its first students were the very first human beings – a mark of its superlatively natural character; this idea, too, Ramus would cleave to in all his later mathematical writings.

In his 1544 oration, then, Ramus adumbrated a number of themes that would occupy a central place in his historiography of mathematics. Yet, in this epideictic oration, he never addressed the question of how mathematics – a frozen snapshot of the real dialectical structure of the human and divine minds and the world – could have a history in the first place. How could change enter such a perfect art? The question must have occurred to Ramus himself, for in his next mathematical work he attempted to address it.

Within a few months of his oration on mathematics, Ramus published a Latin edition of Euclid's *Elements*, his first work devoted to the discipline.[34] This was hardly a great work of scholarship. Ramus printed the Latin text only, without any of the proofs or diagrams – that is, he published only the statements of the propositions. By the standards of the time, Ramus had some justification for his editorial decision to issue such a meager edition. A long tradition maintained that Euclid had not written the proofs, but that the Alexandrian mathematician Theon, who lived half a millennium after Euclid, was responsible for them. Ramus was aware of this tradition (it was, in fact, almost universally accepted in the Renaissance)[35] but he had other reasons for "removing the comments and figures of the interpreters"[36] as he put it in the preface to this edition. On a practical level, omitting them kept the price of the book down,[37] so that it could be used in every school. This was an important consideration, given the place of mathematics in Ramus's educational program: he envisioned the text of Euclid as a key building block of his curricular reform.

Even more to the point, Ramus simply did not see the need for demonstrations in the *Elements*. He referred to demonstrations not as proofs, but as "explanations"; and he suggested that if a student ran into difficulty understanding the material, it would be far easier for his teacher to "explain" it to him in person than for the

---

[33] *Ibid.*, p. 234.

[34] Euclid (1549). The first edition of this work was published in 1545, but only a single mutilated copy survives (Ong, 1958b, p. 34). The edition of 1549 is quite common, and has often been taken to be the first edition; the work was reprinted again, without alteration, in 1558.

[35] On Ramus and the roles and identities of Euclid and Theon, see the sixth chapter.

[36] Ramus (1599, p. 121): "semotis interpretum et commentis et figuris."

[37] *Ibid.*

student to stare at a static diagram.[38] Ramus was still convinced (as he was in the *Institutiones*) that mathematics was an expression of natural dialectic. Here, he had no doubt that the *Elements* was a perfect record of that deep structure of the human mind. In fact, he explicitly identified the propositions of the *Elements* with the "golden chain" of dialectic linking God, human mind and created order, an image he had used for logic in his 1543 works.[39]

But, Ramus believed, as a "natural" science the facts of geometry would be easily and immediately taken up by the student. Therefore Ramus took no interest in the kind of mathematical demonstration found in the *Elements*.[40] To his very last writings on the subject, he would continue to insist that mathematical proof had no place in a genuine science founded upon nature. On the contrary, truths, when placed in their proper, "methodical" relationship with other truths (like the slips of paper drawn from an urn) were *self-evident*; if "explanations" were needed, that in itself indicated a departure from natural order. In the case of the *Elements*, Ramus thought that the proofs were simply superfluous: since Euclid had placed the truths of geometry into their optimal arrangement, the explanations could be omitted without loss of clarity.

As I have said, the problem of the possibility of historical change in mathematics seems first to have occurred to Ramus when he came to write the preface to his *Euclid*, very shortly after the 1544 oration.[41] Here he attempted to account for the problem by presenting a highly idiosyncratic interpretation of the Platonic theory of knowledge. He commended "Plato and Pythagoras" for their notion that mathematics was something divine, lying beyond the human senses; and he echoed an opinion of Proclus that the word "mathesis" meant essentially the same as "reminiscence."[42] The truths of mathematics were, according to Plato, impressed in the mind in imitation of the eternal exemplars of the first intelligence. But, in his gloss on this passage, Ramus did not consider the possibility that each *individual* recalled mathematics, like the slaveboy of the *Meno*. Rather, as he put it,

> What they seem to mean is that so great a science was not invented by man but was divinely impressed in our souls, and by recollection of things that had been noticed was recovered little by little. But how long was that forgetting, and how late a remembering?[43]

---

[38] *Ibid.*: "Si quid autem obscurum fuerit, longe commodius viva praeceptoris intelligentis oratio, quam picta in libris interpretum manus explicabit."

[39] Ramus (1599, p. 120): "Hic enim prima mediis, media postremis, omniaque inter se, velut aurea quadam Homeri catena ... vincta colligataque sunt..." ("For here the first are connected and linked to middles, middles to final like some golden chain of Homer"). For Ramus's use of this image with reference to dialectic, see Bruyère (1984, p. 124).

[40] He does refer to "demonstrationes" at the conclusion of the passage quoted in the previous note, but this refers to his own notion of "demonstration" – synonymous with proper, methodical ordering. From Ramus's point of view, the actual proofs found in the *Elements* were merely elucidations and could be omitted without injury to the mathematical structure of the work.

[41] The preface is dated "5 Cal. Febr. 1544" – i.e., January 28, 1545 (Ramus, 1599, p. 121).

[42] Ramus (1599, p. 120).

[43] *Ibid.*: "quasi tanta scientia non ab homine inventa, sed divinitus in animis nostris impressa, recordatione animadversarum rerum paulatim recrearetur. Verumenimvero quam longa oblivio, quam

Ramus thus turned the Platonic notion of reminiscence into a process in history, and identified it with his own theory of the origins of arts: it was a re-collecting or recording (*recordatio*) of "things that are noticed" – in exactly the same way that dialectic was an art formed by noticing the best practices of natural reasoning. With this in mind, he retold the familiar narrative derived from Josephus, according to which the biblical patriarchs achieved extraordinary advances in the sciences, and were the source of all later Greek learning. Ramus interpreted this well-known story as one of gradual reminiscence or recovery. The first patriarchs, with their preternaturally long lives, could devote decades to observing and recording the mathematical action of the mind.[44] Faced with the impending Flood and concerned about human "forgetfulness,"[45] they inscribed their already substantial results on two pillars, which could withstand destruction by either fire or water. After the re-establishment of the human race, this primitive wisdom was rediscovered and spread through a continuous *translatio studii*: to the Egyptians, Greeks, Italians, Sicilians, Arabs, Spaniards, Germans and, last of all, the French. Countless men, he wrote, have been involved in this "work of recollection" (*recordationis opus*), like so many smiths and architects bringing the edifice of mathematics to perfection.[46]

Although Ramus identified mathematics with dialectic, at this point he invoked two very different historiographical models for the two arts. The story of dialectic was one of gradual discovery, leading up to a moment of perfection and completeness (with Plato), followed by a process of corruption brought on by human pride and arrogance. Mathematics, on the other hand, had developed constantly through "reminiscence" throughout history, though it was not yet complete. Ramus tried to find a compromise between the Josephan record of a *prisca scientia* stretching back to the very first human beings, and the undoubted ingenuity of Greek mathematicians. While mathematics began with Adam and had been transmitted to Ramus's own time in an unbroken chain, the Jewish patriarchs were not doing anything *qualitatively* different from any other mathematician involved in the great act of remembering. It seems that Ramus would reject Josephus's notion that Adam and the first humans had a superlative knowledge of mathematics which we can scarcely approach. Instead, Ramus clearly valued the Greek achievement: it was the *Elements*, after all, which had gathered together all the isolated mathematical facts from earlier practitioners and – like the grammarian picking slips of paper out of an urn – placed them all in just the *right* places.

---

tarda recordatio ista fuit?" For other examples of Ramus's assimilation of Platonic metaphysics to his own theory of knowledge, see Bruyère (1984, pp. 262–264).

[44] Ramus (1599, p. 120): "Primi illi homines (ut Josephus antiquitatis Judaicae scriptor ait) Adamus, Sethus, Enus, Noeus vitae et longissimae et contemplationi deditissimae beneficio, in hanc recordationem incubuerunt." It is worth noting that Ramus included Adam in this list, even though Josephus did not explicitly say that he had pursued mathematics.

[45] *Ibid.*: "ne alia novae oblivionis caligine circumfusa teneretur..."

[46] Ramus (1599, pp. 120–121): "Hinc tot, tamque excellentia ingenia excitari ... coeperunt, videlicet ad huius mathematicae recordationis opus exaedificandum, tot fabros, tot architectos adhiberi oportuit..."

A decade later, however, when he came to write his next mathematical work, Ramus reached very different conclusions. In his *Arithmetica* of 1555, he provided mathematics with a history that looked much more like that of dialectic, complete with villainous and selfish corrupters of the arts. What had happened? It seems that, in the intervening years, Ramus had devoted some time to the actual study of mathematics, and had discovered that the art was not at all what he had assumed it to be. When he published his *Euclid* in 1545, he clearly thought that mathematics was the most natural and well-organized of arts; therefore, it would be easy to learn. In his intellectual apology *Oratio de professione sua* written in 1563, Ramus recalled that he later had a change of heart, probably some time in the years 1551–1555[47]:

> There are 15 books of Euclid's *Elements* which I thought had been put together by the one and only instrument of Logic – just like absolutely every other art. Thus, I thought, it could subsequently be analyzed by means of the same instrument. In fact, I had long since devoted myself to logic, preparing it for the sake of mathematics above all else. I was persuaded by my own argument and tried to ignore the many great obscurities endemic to mathematics; by hard work and my own sharp mind I got all the way through to the 10th book. Pierre de Mondoré[48] had been most eruditely explaining and clarifying that book for me. However, its immense subtlety still exercised me enormously. One day ... in fact, I had been trying unsuccessfully to get to the end of a demonstration on the binomial and residue; I concentrated my mind on it entirely; after keeping my body stuck in one position for a whole hour, I felt all the muscles in my back seize up. And at that, I threw away my drawing-board and ruler, and burst out in rage against mathematics, because it tortures so cruelly those who love it and are eager for it.[49]

Quite against his expectations, mathematics – the goal and paragon of his natural, logical method – turned out to be *hard*.

This defeat marked a major crisis in Ramus's intellectual development. He had built an entire philosophy on the association of natural dialectic with mathematics. But now it turned out that mathematics was *not* a natural, immediately graspable science at all. To remedy this situation, Ramus did two things: first, he wrote his own mathematics (in the form of his *Arithmetic*, and later his *Geometry*, *Algebra* and

[47] In the 1563 oration, when relating the incident described below, Ramus said that he had spent 8 years of his career teaching the trivium, followed by four years of teaching mathematics. If he is dating the beginning of his career from his publications of 1543, then the mathematical phase of his career (as he considered it) was 1551–1555.

[48] Mondoré (1552).

[49] Ramus (1599, p. 409): "Quindecim Euclidis libri sunt, quos (ut omnes omnino artes) sicut uno Logicae organo contextos esse primum, sic eodem postea retexi posse cogitabam. Organum autem illud una imprimis mathematum causa diu multumque praecultum nobis ac praeparatum est. Quare persuasione hac inductus nihil reputans quot et quantae mathematum per se obscuritates essent, prompto atque alacri animo ad decimum usque librum penetravi, sed immensa subtilitate operis illius, licet eruditissimis P. Montaurei vigiliis explicati et illustrati, tamen sic exercitatus sum ... ut quodam die cum binomii et residui cuiusdam demonstrationem summa animi intentione, corpore horam integram idem vestigium premente nondum conclusissem, senserim collo nervos obriguisse: tum vero abacum radiumque abieci, indignatusque mathematis succensui, quod sui studiosos et amatores tam acerbe cruciarent." See also Waddington (1855, p. 108).

*Optics*) which would form the core of his own mathematical teaching; and second, he returned to writing the history of mathematics, this time in far greater detail.

In the preface to his 1555 *Arithmetica*, Ramus again presented the standard narrative of the origins of mathematics, just as he had done in the preface to his *Euclid*. Adam, Seth and Noah had spent their time contemplating mathematics "in order to appreciate God's mathematical work."[50] Abraham passed the science on to the Egyptians, whence it was taken up by the classical civilizations. In ancient Greece and even, to some extent, in ancient Rome, there followed a golden age of mathematics. Boys studied the art, practicing it by drawing in the sand, and craftsmen like painters and architects both knew and used mathematical techniques.[51] In other words, an art of mathematics existed, but one that conformed to natural mathematics, as it was expressed by human beings using arithmetic and geometry to practical ends. Ramus was still thinking about the relationship between mathematics and dialectic; now, however, rather than seeing it as the lofty pinnacle of dialectic, he had the eloquent vine-dressers of the 1543 *Institutiones* in mind, and the *useful* dialectic he intended to teach in the schools.

What happened after this golden age? There was a collapse into barbarism for many centuries, followed by a revival of mathematics, but only in a limited sense: those who revived the mathematical arts in Europe thought "that they were not like the others, which were useful after they had been learnt, but only *while* they were being learnt"[52] – that is, as a tool to sharpen the mind before going on to "higher" disciplines, such as philosophy. The very difficulty of mathematics thus became a recommendation for its study – for Ramus an absurd, even self-contradictory idea. And what was the cause, both of mathematics' precipitous decline and later, its perverse revival? More than anything, Ramus said, it was the obscurity of the subject. He laid the blame for this state of affairs squarely at the feet of Euclid and Theon.

By Ramus's lights, if Euclid really *had* put together the *Elements* according to the natural method, then his work should pass the test of the three "laws of method" which Ramus developed after his return to philosophical publication in 1551.[53] Not surprisingly, Euclid failed miserably as a Ramist logician. Arithmetical precepts were (said Ramus) often expressed in terms of general magnitude, which was properly the province of geometry. This broke the "law of justice" or homogeneity: "arithmetic should be taught arithmetically, geometry geometrically."[54] Further, number should be logically prior to magnitude, according to Ramus at least; yet the *Elements* began with geometry – clearly a violation of the "law of wisdom," which required more general sciences to precede more particular. Lastly, Euclid

[50] Ramus (1599, p. 121): "Haec enim primorum generis humani parentum, Adami, Sethi, Noëi divina contemplandis optimi maximique Dei mathematicis operibus otia fuerunt."

[51] Ramus (1599, p. 122).

[52] Ramus (1599, p. 122): "sed [opinantur] prodesse has artes non caeterarum more, cum perceptae fuerint, sed cum percipiuntur."

[53] See Ong (1958a, pp. 258–262).

[54] Ramus (1599, p. 123): "Itaque arithmetica arithmetice, geometrica geometrice doceantur."

introduced his definitions in two groups, at the beginning of books I and V; but as the grammarian with an urn of grammatical facts knew, precepts of an art should be introduced only where they belong in the natural order of things.[55]

Ramus concluded that Euclid, although no doubt a fine collector of individually excellent mathematical truths (most of them unearthed long before him by the first human beings), was a dunce when it came to arranging them according to their nature. In another departure from his position in the 1545 *Euclid*, Ramus now believed that the demonstrations were a necessary part of the *Elements*, and that this was itself another black mark against Euclid's name. Euclid may have written his work without demonstrations, thinking it was sufficiently clear in that form; but the very fact that Theon thought it necessary to add explanations to the originally naked text only confirmed – and deepened – the obscurity of Euclid's arrangement.[56]

Ramus required a properly arranged mathematics, "by which an absolute beginner who wants a perfect and complete grasp of the art can be perfectly and completely taught"[57] – a *natural* mathematics, in which there would be no need for demonstrations. He exhorted his readers:

> Establish, finally, the elements of mathematics according to these laws of logic: the individual propositions arranged in place and order will not only be statements of their own truth, but even *demonstrations* of it.[58]

Ramus tried to provide precisely this in his *Arithmetica* and other mathematical books; although he was never completely satisfied with his reformed mathematics, they remained popular school textbooks for more than a century after his death.[59]

Ramus had now constructed a history of mathematics which in outline essentially matched his history of dialectic: an original, natural state, corrupted by the pride and arrogance of Euclid who, like Aristotle, elevated his "contrived demonstration" over the natural structure of discourse.[60] Ramus's narrative was also beginning to take on the shape of the story of sacred history itself: innocence, fall and finally redemption (through the application of proper method). In the final expression of his thought on the question, the *Prooemium mathematicum* of 1567, Ramus would

---

55 Ramus (1599, pp. 123–124).

56 Ramus (1599, pp. 124–126): "Atque haec elementa licet a primis usque hominibus repetita, tamen hunc in modum et proposita et collocata ab Euclide existimantur; quem virum mathematica singularum propositionum scientia tanquam singularem et prope divinum suscipio ... at logica recte et ex ordine docendi prudentia parem efficere nequeo. ... Quare licet [Theonem] mathematicarum rerum intelligentia non inferiorem putemus, attamen videmus adhuc quam demonstrationibus suis elementa mathematica non illustret, sed obscuret."

57 Ramus (1599, p. 123): "quibus perfecte et absolute rudis et imperitus institui possit."

58 Ramus (1599, pp. 126–127): "Denique mathematica elementa logicis legibus illis institue; propositiones singulae loco et ordine collocatae, ipsaemet suae veritatis non tantum propositiones, sed etiam demonstrationes erunt."

59 See Ong (1974), for Ramus's continuous emendation of his mathematical texts.

60 Ramus (1599, p. 126): "non Aristotelis commentitia illa quidem, sed certe naturali et aperta demonstratione singulariter et eximie demonstrari."

make explicit the connection between disciplinary and religious history, while substantially modifying in every detail the history of mathematics as he had received it from earlier authors, and even as he himself had presented it in his series of mathematical prefaces.

# Chapter 3
# From Plato to Pythagoras: The *Scholae mathematicae*

## Introduction

Ramus's great contribution to the history of mathematics, the *Prooemium mathematicum*, was written in a time of confessional strife, both in France and in Ramus's own life. Some time in 1561 or 1562, Ramus converted to the Reformed religion – to no one's surprise, since many had long suspected that he had secretly embraced Protestantism. In 1562, when Calvinists were expelled from Paris, Ramus – by then one of the most famous scholars in the world – was given royal safe-passage to Fontainebleau.[1] He worked in the library there for several months, reworking the lectures on mathematics that he had developed over several years. These mature thoughts on mathematics and history would eventually be published in two versions: the *Prooemium mathematicum* of 1567, and the *Scholae mathematicae* of 1569, which contained the three books of the *Prooemium* scarcely altered, plus another 28 books of criticism of Euclid, extending his brief remarks of the 1555 *Arithmetic* in exhausting detail.[2] I will argue in this chapter that Ramus's elaborate reworking of his history of mathematics bears witness to the religious and civil strife that overtook France in those years, as well as to the academic conflicts in which Ramus found himself embroiled in the University of Paris and the Collège Royale.

In the previous chapter, I reconstructed Ramus's history of mathematics as it developed in the prefaces he wrote to various early mathematical publications. The history of mathematics in the *Prooemium* of 1567 represents Ramus's final revision of this narrative. Like the earlier versions out of which it grew, it is a *critical* history; throughout, Ramus assessed the mathematical past, assigning praise and blame to its actors as he proceeded – his aim, as ever, to communicate a model of mathematical education and practice through the historical narrative he so deliberately constructed.

In his mathematical prefaces, Ramus had constantly promoted the idea of a "natural" mathematics. With his 1555 preface, he launched his offensive against

[1] Waddington (1855, pp. 136, 149–150).

[2] Ramus (1567) and Ramus (1569). In what follows on the *Prooemium mathematicum*, all references will be to the more widely-accessible *Scholae mathematicae*.

R. Goulding, *Defending Hypatia*, Archimedes 25,
DOI 10.1007/978-90-481-3542-4_3, 

Euclid and Theon, whose logical ineptitude had veiled this natural mathematics in obscurity. Later, on his return to Paris from his 1562 exile, he declared that the months of enforced isolation had enabled him to sharpen up the attack on the "futile subtlety" of Euclidean mathematics.[3] In the *Prooemium mathematicum*, Ramus would elaborate upon the charges made against Euclid and Theon in extraordinary detail. And, while leaving no doubt of their culpability, Ramus addressed an omission of the earlier work: there, the willful destructiveness of Euclid and Theon seemed almost unmotivated. In the *Prooemium mathematicum*, Ramus provided a sustained historical account of the reasons for their errors. By doing so, he hoped to rescue modern mathematics from its twin defects of inutility and obscurity. Pinpointing the moment when mathematics went astray, Ramus also discerned the genuine, primitive purity of the art hidden under the surface of human artifice.

In all three books of the *Prooemium*, Ramus dealt with the history of mathematics in depth, but in different ways and to different ends. The first book concerned the history of ancient mathematics, beginning with biblical and mythical prehistory and finishing with Theon, editor of the *Elements*, whom Ramus took as the end-point of ancient science. In this book, Ramus laid down his fundamental interpretive principle for the history of mathematics, which he would apply throughout the rest of the *Prooemium* and *Scholae mathematicae*: that mathematics had gone into decline whenever its practitioners succumbed to the temptations of elitism and abstraction. It had flourished, on the other hand, whenever it remained open to artisans, directed towards application, and founded upon nature.

In the second book, Ramus defended mathematics against charges of uselessness. As one might expect, he catalogued here the many practical applications of mathematics, as well as its utility in all the other arts, from natural philosophy to theology. But he also resumed his historical narrative – this time with a very particular geographical focus: having finished the previous book with Theon of Alexandria, he took up the story again with Regiomontanus, devoting the rest of book II entirely to modern German mathematicians. As he presented them, these practitioners had embraced applied mathematics, eschewing theoretical flights of fancy. Despite their practical orientation (or, rather, because of it) they had no difficulties finding either patronage or, most significantly for Ramus, employment in the universities. In their success, the Germans were different from practitioners in England (who, as Ramus lamented in the first book of the *Prooemium*, received little encouragement of any kind) and in Italy (the subject of his historical excursus in the third book). But the most glaring contrast Ramus drew was with the contemporary University of Paris. Throughout the *Prooemium*, Ramus flung barbs at his own institution; his survey of the successes of German mathematics allowed him to set out an extended critique of Paris and its neglect of mathematics. To heighten the contrast with the heroic German practitioners, Ramus personified the errors and idiocies of Paris in the character of "Aristippus" who, as in Melanchthon's encomium on mathematics, stood for the hater of the science who belatedly discovered its worth. In Ramus's

[3] Ramus (1599, p. 410).

version, however, Aristippus was also a thinly veiled caricature of his nemesis at the University, Jacques Charpentier. In fact, Ramus reworked the entire historical narrative in the *Prooemium* in response to his very public loss to Charpentier in a court case concerning the future of mathematics at Paris.

In the third and final book of the *Prooemium*, Ramus tackled the next charge after "uselessness" that he said was often laid against mathematics: that of "obscurity." Here Ramus returned to the attack on the *Elements* and the errors of Euclid and Theon that he had begun in his prefaces. Now, however, he was able to rest his critique on solid historical grounds (or so he thought). In the new material that he added to the *Prooemium* in the 1569 *Scholae mathematicae*, Ramus continued in this critical vein. After a brief précis of his own *Arithmetic* in books 4–5, Ramus devoted the remainder of the *Scholae* (books 6–31, 175 dense quarto pages) to a definition-by-definition, theorem-by-theorem critique of the *Elements*.

In both the *Prooemium mathematicum* and the expanded *Scholae mathematicae*, Ramus endeavored to show not only that mathematics was the most ancient of all arts, but also that from the earliest date it had been central to a liberal education in *precisely* the way he had long argued it ought to serve the modern academy. In constructing this argument, Ramus had very limited resources to draw upon. He was concerned to extract from the fragmentary narrative of antiquity a continuous story of mathematics, from its most primitive beginnings to historical times, at times pushing the evidence far beyond where it reasonably led. Through a combination of solid research, wishful thinking and, it must be admitted, occasional falsification, Ramus constructed a coherent narrative of mathematics' past that supported his contemporary educational program: mathematics in its formative beginnings (and hence in its essential nature) looked very much like the reformed mathematics he wished to have taught at the University of Paris.

Both Ramus and his opponent Charpentier used and, sometimes, abused the history of mathematics in order to make larger points in their pedagogical and ideological disputes. Examining just how they did so can provide insights into the role of history and the historical imagination in the development of the sciences in this period. The debate with Charpentier turned in part upon the utility or inutility of mathematics. Ramus would devote a whole book of his *Prooemium* to refuting the charge that mathematics was "useless." In fact, this term did not correspond at all what we might expect it to mean. The problem was not that mathematics was useless, but rather that it seemed *too* useful, the tool of navigators, merchants, surveyors and builders – not at all the sorts of professions to which the classes attending the university aspired. Imagining a past for mathematics amounted to imagining a world in which mathematics had a place; a representation of the ancient world as a place where mathematics and its applications were studied, taught and valued could provide a useful model for contemporary practice. And such a model gained in significance the further back into the past it could be constructed: the beginning of mathematics was the surest guide to what it *ought* to be.

In the course of this chapter, I shall examine the role above all of two principal historical figures in Ramus's history: Plato and Pythagoras. The latter assumed an importance in the *Prooemium mathematicum* that he had not had in any of Ramus's earlier mathematical writings – a prominence, I argue, that arose from the clashes between Ramus and Charpentier. Plato, on the other hand, had been respected by Ramus in all his earlier writings; in the *Prooemium*, he is portrayed in a much more uncertain light. First, however, I shall take up the same subject that interested Ramus in his mathematical prefaces: the origin of mathematics.

## The Origins of Mathematics

From the very beginning of the *Prooemium mathematicum*, Ramus sounded a theme that was quite familiar to readers of his various mathematical prefaces. Aristotle, reported Ramus, believed that the arts were eternal, like the world itself – yet, like the stars, their fortunes rose and set through history. Ramus agreed enthusiastically: "A truly great saying of a great philosopher: the arts are of eternal and unchanging things, yet men's knowledge of them is by no means eternal."[4] Committed as he was to the correspondence between the *axiomata* of the arts and the facts of the world, whether taken individually or in organized groups, Ramus would agree that the arts were as eternal as the world they mirrored. But human beings regularly distorted or forgot the natural arts that they had once possessed. The story that unfolds in the *Prooemium* is, in part, an account of the fortunes of eternal arts and ideas in the hands of limited and fallible mathematicians.

Ramus opened the book with a rapid survey of the earliest origins of mathematics (or, at least, of astronomy, for which more evidence was available). The structure of his narrative followed Pliny's division of astronomy into four "sects:" Chaldean, Egyptian, Greek and Roman. But whereas Pliny had been interested in the different *schools* of geometry and their divergent beliefs, Ramus emphasized that these were *periods* in the development of astronomy. *Commigrationes disciplinarum* – the migration of learning – mirrored the migration of peoples (*commigrationes gentium*), as Ramus adapted Pliny's observation on the styles of astronomical practice to conform with the well-established narrative of the history of mathematics, passing from people to people, from "barbarians" to the ancestors of European civilization.[5]

Ramus pushed back Pliny's Chaldean "period" into the age of the Hebrew patriarchs. By citing Berosus (as he was attested in Josephus)[6] Ramus was able to make this and the subsequent Egyptian age congrue with Scriptural chronology, importing into his narrative not only the transitional figure of Abraham but the whole

[4] Ramus (1569, p. 1): "Haec magni philosophi magna prorsus sententia est, artes sunt aeternarum et immutabilium rerum, at ipsarum apud homines notitia nequaquam est aeterna."

[5] *Ibid.*; see Pliny, *Natural History* XVIII.211.

[6] *Pace* Popper (2006, p. 96), there does not seem to be any influence of Annius of Viterbo in Ramus's *Prooemium mathematicum*. While Ramus was undoubtedly aware of Annius's fictions (which played an important role in his work on the ancient Gauls) all of Ramus's references to Berosus in the *Prooemium* are to the *genuine* Berosus cited by Josephus, not the Annian pseudo-Berosus.

Josephan apparatus of patriarchal mathematics: the long-lived patriarchs, the pillars of Seth, Abraham and the "migration of mathematics from the Chaldeans to Egypt."[7] Ramus quoted all of this from Josephus with frequent approbation, occasionally adding corroborating details (for instance, that, Berosus himself was honored by the Athenians for his astrological expertise, thereby underlining the connection between the Chaldeans and ancient astronomy).[8]

As he drew closer to the historical age, Ramus began to weave in Proclus's quite different account of the beginnings of mathematics. According to Proclus, geometry began with the Egyptians themselves, who needed to restore land boundaries after the flooding of the Nile. In answer to this, Ramus agreed that the Egyptians during the imperial age or, even earlier, from the time of the foundation of Alexandria by Alexander the Great, were renowned for their skills in practical mathematics. This was, he implied, the source of Proclus's erroneous assumption that mathematics had its origins in Egypt: his knowledge was limited to Greek reports of Egyptian proficiency.[9] Sacred sources, however, showed that the knowledge the Egyptians drew upon to restore their fields had been given to them many centuries earlier, long before Alexander made Egypt a part of the Greek world. Indeed, Aristotle's claim that Egyptian priests discovered mathematics during their abundant leisure was also corroborated by Scripture. Genesis records (said Ramus) that Joseph, under Pharaoh's orders, bought up all the land of the Egyptians during the famine, but that "Pharaoh granted the priests their land as a stipend for their profession of mathematics."[10] The Bible in fact says only that "the priests had a fixed allowance from Pharaoh, and lived on the allowance which Pharaoh gave them; therefore they did not sell their land."[11] Not even Josephus connected this royal stipend with mathematics;[12] but for Ramus, it made perfect sense to harmonize all of his sources in this way. The priests, pursuing in their free time the mathematical knowledge they had inherited from the Hebrews, mirrored the ancient patriarchs themselves, who pursued the sciences in the protracted stretches of their God-given lifespans. The conflation of anecdotes also allowed Ramus to make a point for the first time which he would reiterate frequently in the *Prooemium mathematicum*: that mathematics deserved royal patronage, indeed, had always been an art patronized by kings. The achievements of the Egyptian priests, he concluded, were "monuments to the kings;" they only constructed their marvelous mathematical instruments and schools through royal generosity.

In the early pages of the *Prooemium*, Ramus concentrated on tracing the origins of astronomy from the remotest antiquity. He said little about the ultimate origins

---

[7] Ramus (1569, pp. 2–4).

[8] *Ibid.*, p. 3. Ramus discovered this fact in Pliny, *Natural History* VII.123.

[9] *Ibid.*, p. 4.

[10] Ramus (1569, p. 4): "ut a rege Pharaone sacerdotibus ager in stipendium mathematicae professionis esset assignatus."

[11] Genesis 47:20–22. Ramus incorrectly cited the passage as Genesis 27.

[12] In *Jewish Antiquities* II.189, he simply says that Pharaoh took possession of all the land of Egypt "save only the priests, for these kept their domains."

of geometry or arithmetic, but was confident that they, too, must have been developed at an early stage: "arithmetic and geometry must necessarily have preceded [astronomy],"[13] he said, since these sciences were necessary for the practice of astronomy – an argument that he would repeat several times in the *Scholae mathematicae*. As he did in his mathematical prefaces, Ramus guaranteed an origin for all the mathematical sciences that predated the Flood. Nowhere in this survey of the early history of mathematics did Ramus comment on the *means* by which mathematics had been discovered. In the mathematical prefaces, he had built up a model of mathematical discovery based on the conformity between the mind in its natural state and a certain kind of "natural mathematics." Like the dialectic for which Ramus was so famous, the *art* of mathematics was secondary to its untutored, primitive practice: over generations, skilled observers had recorded the activity of its best practitioners, forming a progressively more complete documentation of the native art (a process Ramus called "recollection"). In the much more detailed history of early mathematics in the *Prooemium*, however, Ramus had nothing to say on the process of discovery.

It will be my argument that, despite his silence here, Ramus retained a concern in the *Prooemium mathematicum* for natural mathematics and its corruption; that, indeed, one of his central concerns in the *Prooemium* was to explain just how mathematics had departed from its natural state. But Ramus's opinion on how mathematics arose is buried quite inconspicuously in the text. Towards the end of the third book of his lectures, Ramus surveyed the recent mathematical accomplishments of all the nations of Europe, especially Italy, exhorting the rulers of each state or city to encourage the study of mathematics in their schools. When he came to discuss Rome, the apostate Ramus found himself in the awkward position of having to address the pope. He did so by turning the austere Pius V into a kind of pontifical Protestant. In his simplicity of life (wrote Ramus), the pope recalled the early Church Fathers; surely, then, he would restore a new, primitive Christianity to the Catholic Church. Such an ecclesial restoration, Ramus went on, should be accompanied by a reformation of the arts. The first humans, he explained, did not have to learn grammar because they all spoke the same language. Nor did they have any need for rhetoric, since there was no disagreement at all in that blessed state. As for logic, their reason was utterly unblemished, and they used their natural faculties without error. "And so (he says) the mathematical arts [alone] were divinely given or discovered, so as to reveal the power of God in the creation of the world, his wisdom in its administration, and his faithfulness in providing so many benefits to the human race."[14] Therefore he urged the pope not only to restore primitive religion, but also to support the teaching of man's first knowledge in the schools, to the end of healing the rifts in Christendom.

---

[13] Ramus (1569, p. 4).

[14] Ramus (1569, p. 109): "Ergo artes mathematicae divinitus vel oblatae vel inventae, quae Dei potentiam in mundi creatione, sapientiam in administratione, pietatem ex infinita bonorum omnium erga genus humanum largitate demonstrarent."

In this passage, Ramus seemed to have taken a quite different position on the origin of mathematics than the one that he had staked out in his prefaces. Before, he had thought of mathematics and dialectic together: both of them paradigmatic natural arts, born out of the observation of human practice. Dialectic *as an art* arose only later, as human beings discerned a need to record and teach the best examples of natural reasoning. By analogy, one would expect (as in the prefaces) early humans to have practiced mathematics purely according to the light of reason, only later codifying it into an art. But in this context Ramus separated mathematics from dialectic. The art itself of mathematics (not just the natural human affinity for it) was given to men by God, so that they could admire his handiwork and praise him worthily.

Under this conception of the science, there was not much room for growth or even the "reminiscence" which had played such a central role in Ramus's earliest history of mathematics. The heroes of the *Prooemium* would be those who imposed order on the art, or who extended its range of application. The material of the art was given by God to the very first human beings, while its form was the work of practitioners who understood the natural working of reason.

The new conception of the origins of mathematics which Ramus put forward in this passage went hand in hand with his concern during this period for the settlement of religious disputes. In his posthumously published *De religione christiana*, Ramus used very similar language about theology. There, he stated that theology, the "art of living well," consisted of the methodical arrangement of Biblical verses. Any religious disagreement could be resolved by consulting a properly organized collection of scriptural quotations. Just like mathematics, the material of theology (Scripture) had been given by God, while the form had to be discovered by human effort, by placing the individual sayings of Scripture into the single possible logical and natural structure (as Ramus thought). Once this had been done, the practice of theology – like any other "methodized" art – would be easy and open to all human beings.[15]

It is quite appropriate, then, that Ramus would set out his view of the origins of mathematics in an address to the pope, calling for a curricular reformation that would also heal the divides in Christendom. Such a conception of mathematics further explains Ramus's insistence in the *Prooemium mathematicum* on *commigrationes disciplinarum*, and on drawing a continuous lineage from the earliest human beings to the present day – even to the extent of having to explain away the independent discovery of geometry by the Egyptians. Of course, the *translatio studii* also featured in his earlier histories; but in the *Prooemium* the phenomenon assumed a much more central role. If God had delivered the contents of the art itself to the first human beings, it was important that there should be no breaks in the line of transmission.

This is not to say that Ramus in any way abandoned his commitment to the notion of natural mathematics, nor his belief in the affinity between the human mind and a properly ordered science – even a natural *practice* of mathematics. In one of

[15] Ramus (1576, p. 2).

the most vivid passages in the *Prooemium*, Ramus took his readers on a virtual stroll through the streets of Paris, observing the merchants, craftsmen and lawyers unselfconsciously using mathematics in the course of their ordinary business.[16] The ubiquity of arithmetic and geometry demonstrated not only that these arts were *useful* (his immediate concern in this part of the *Scholae*) but also that they grew out of the natural mental faculties of counting and measuring. Ramus quite categorically rejected the Aristotelian conception of the mind as a "tabula nuda."[17] Nevertheless, there was a definite shift in emphasis in the *Prooemium*. His revised view of the process by which the developed arts were formed stemmed in part from his contemporary religious concerns. It may also reflect a new pessimism in his thinking: as he became more deeply versed in Greek mathematics, Ramus had perhaps begun to doubt that the most advanced results could have been discovered by the mere human ingenuity of the first patriarchs (a disillusionment that mirrored his earlier frustration with the *Elements*).[18]

## Ramus and Plato

In the mathematical prefaces, Ramus included Plato among the pioneers of Greek mathematics. Moreover, as I argued in the last chapter, the model of mathematical progress that Ramus proposed in those texts was an idiosyncratic kind of Platonism. In the *Prooemium mathematicum*, however, Ramus's attitude towards Plato was more complicated. Socrates remained an untarnished hero, as he was in so many of Ramus's writings.[19] Indeed, Ramus identified himself with Socrates (or Socrates with himself) to a remarkable degree. In his most extended remarks in the *Prooemium* on the ancient philosopher, Ramus explained that the chief genius of Socrates was to make the liberal arts answer to human needs. Socrates thought that there was too much reading of books in the schools, too many arguments over empty trifles. The way to make sailors, architects and farmers was not to argue over the precepts of their arts, but, once those arts had been quickly understood (*breviter intellectae*), they should be learned by *doing*. Ramus maintained that he had learnt this "Socratic philosophy" from Plato and Xenophon, and that he had suffered just the same fate as his idol, except that he had not been condemned to death (a rather large exception, one might say):

> When I entered the University of Paris, I suffered the miserable and astounding ill-fortune of Socrates. By everyone's judgement, I was condemned as an offensive and ignorant slanderer, even as an enemy of religion (for some thought that religion was founded on scholastic sophisms); and I was prohibited from writing or speaking publicly or privately.

---

[16] Ramus (1569, pp. 54–55). Much of this passage is taken from one of Ramus's mathematical *actiones* against Jacques Charpentier, discussed below.

[17] *Ibid.*, pp. 82–83.

[18] See the previous chapter, at p. 30.

[19] On Ramus's "socratism," see Walton (1970).

> This was like cutting off my hand and tongue. Apart from the hemlock, I went through everything that Socrates did.[20]

It was only when Henri III restored his right to teach and elevated him to a regius chair that Ramus was free to proclaim the "Socratic philosophy" again.

At times, Ramus was able to praise Plato just as highly. In his exposition of the transmission of mathematics from the barbarian world to the Greeks, Plato played a crucial role. According to Ramus's account in the *Prooemium*, the original mathematics of the patriarchs was further developed by three separate schools, each of which shaped the science according to its particular interests. The Egyptians were, of course, the oldest of the Mediterranean peoples to practice mathematics. In the Greek world, the "parents" of the art were Thales, who brought mathematics from the Egyptians to the Greek mainland, and Pythagoras, who established a mathematical school in southern Italy.[21] It was Plato, wrote Ramus, who reunited all three traditions: he studied with Theodorus Cyreneus in Greece, with the Pythagoreans in Italy, and with the Egyptian priests themselves. Having brought together the divergent strands of mathematics, Plato made many discoveries himself and, influenced by Pythagoras, recognized the importance of mathematical "elements" in the education of the young.[22]

All this made Plato the "prince of mathematicians,"[23] and Ramus could identify himself in this role as readily as he could see himself as a new Socrates. This can be seen most clearly in Ramus's treatment of the "Delian problem," a well-known story that concerned a plague afflicting the people of Delos. Told by an oracle that they needed to double the size of their cubical altar to Apollo – without, however, changing its cubical shape – they turned in bewilderment to Plato. After admonishing them and their fellow Greeks for their ignorance of geometry, Plato passed their problem on to one of his friends. According to the most familiar version of this story (recorded by Valerius Maximus), that friend was none other than Euclid himself. (This anecdote and its influence will be examined closely below, in fifth chapter, in the context of humanist biographies of Euclid.) For Ramus too, the anecdote was important for establishing the identity of the author of the *Elements*; but it was equally significant for what it said about *Plato*. The Delian problem and Plato's

---

[20] Ramus (1569, p. 77): "cum ... in Academiam Parisiensem induxissem, Socratis miseriam et calamitatem mirabilem mihi conciliavi, judiciis omnibus, pro impudente et ignaro calumniatore, pro impio etiam damnatus; religionis enim fundamentum nonnulli tum in scholasticis sophismatis collocarant; scribere quicquam aut loqui publice privatimque prohibitus. Id fuit manus et linguam velut amputare; denique Socratis praeter cicutam nihil nobis admodum abfuit."

[21] *Ibid.*, pp. 5–6. It is a little unusual that Ramus said nothing here of Pythagoras's own travels in Egypt, as so many other historians did – even Ramus himself, who mentioned Pythagoras's journey to Egypt in his second *actio* against Charpentier, written the previous year. See Ramus (1599, p. 419). Rather, he gave the impression that Pythagoras was a linear successor of Thales.

[22] *Ibid.*, p. 12: "Sed unus mathematicorum omnium, tanquam Homerus, habetur Plato, qui non solum a Theodoro Cyreneo in Graecia, a Pythagoreis in Italia, a sacerdotibus in Aegypto mathematica tum inventa didicit, sed per sese multa exprompsit." Plato's supposed admiration of Pythagoras's "elements" will be explained below.

[23] *Ibid.*, p. 15.

role in its solution was, in fact, one of the recurring motifs of the first book of the *Prooemium*. Through variation and repetition of this anecdote, Ramus molded the figure of Plato until the ancient philosopher began to look very much like Ramus himself.

The episode in Delos formed the centerpiece of Ramus's life of Plato, which itself appeared in the context of an exhortation to the University of Paris (one of many in the *Prooemium*). Plato and Pythagoras, Ramus wrote, desired the University to embrace the study of mathematics; for both had placed mathematics at the very foundations of a liberal education.[24] In order to underscore the weight of Plato's opinion about the study of mathematics among his contemporaries, Ramus related the plight of the people of Delos.

Plato was held in such high regard in his own day, Ramus said, that the Delians turned to him *tanquam mathematicorum principem*, "as prince of mathematicians" to solve their problem.[25] Ramus did not cite the familiar account of Valerius Maximus, but a text that would be obscure for anyone other than a logician like himself: John Philoponus's commentary on the *Posterior Analytics*. That text reads as follows:

> "Let two cubes become one cube" means "how is it possible to double a cube while it yet remains a cube?" The following well-known story seems relevant to this. When the Delians were suffering from the plague, Apollo told them that they would be freed from it if they could double an altar which was in the shape of a cube. So they built another, identical cubical altar and put it next to the first. But the juxtaposition of two cubes changed the shape of the altar: instead of a cube, it was now a rectangular prism. When the plague did not cease, the god answered that they had not done what was ordered. ... So they came to Plato, asking how the cube could be doubled. He told them, "The god is reproaching you for your neglect of geometry. For the duplication of the cube will be solved, when two mean proportionals to two straight lines are found." This latter problem he proposed to his students, each of whom wrote about it according to his ability. But nothing that they wrote was of any use. Nor did the Geometer indicate a solution ..."[26]

As Philoponus went on to explain, "the geometer" (that is, Euclid) showed only that the duplication of the cube was equivalent to the discovery of two mean proportionals; he did not show how to find the proportionals themselves.[27] That problem

---

[24] Ramus (1569, p. 13).

[25] *Ibid.*

[26] Philoponus (1542, p. 36) (commentary to *Post. An.* I. 7): "Quod autem duo cubi si[n]t unus cubus, hoc dicit, quomodo possibile est cubum duplicare et rursus manere cubi figuram, videtur autem pro hoc vulgata innui historia. Daliis [*sic*] enim peste laborantibus, respondit Apollo fore ut liberarentur a peste, si aram duplicarent cubicam habentem formam, hi autem edificarunt addentes priori arae, alterum cubum aequalem. At duorum cuborum compositio cubi formam permutavit, fuit enim pro cubo trabs. Peste autem non cessante, Respondit dominus, non fecisse eos quod praeceptum fuerat. ... Venerunt autem ad Platonem quaerentes viam quomodo cubum duplicarent, hic autem ad ipsos ait, videtur vobis improperare dominus, veluti negligentibus geometriae. Duplicatio enim cubi invenietur inquit, si duarum rectarum duae mediae proportionaliter inveniantur, et hoc problema discipulis proposuit, qui de hoc scripserunt ut potuit unusquisque quorum nihil servatur usque modo, sed neque geometra hoc significavit ..."

[27] Euclid, *Elements* XI.33 and corollary.

was solved by later mathematicians, starting with Apollonius, whose solution he presents in detail.

In the *Prooemium mathematicum*, Ramus summarized Philoponus's version of the anecdote accurately enough, at least up to the point where Plato interpreted the request as an admonition from the god. But, in Ramus's retelling, Plato did not simply hand over the problem to some of his students for solution, as he did in Philoponus and every other version of the story. According to Ramus,

> Plato immediately sent out letters to all of his friends – letters to Italy, to Egypt and to the whole of Greece – urging all the accomplished geometers to solve this problem.[28]

In Ramus's imagination, Plato's letter-writing campaign had provoked a flurry of mathematical activity across the civilized world.[29] Ramus's model for this quite creative reimagining of Plato was, of course, himself. His goal throughout the *Prooemium* was to revive the study of mathematics, not only at the University of Paris, but in all the universities of Europe. In preparation for this task, he had sent letters to every corner of Europe, exhorting mathematicians to foster the study of their arts in their own countries or urging them to work on intractable mathematical problems.[30]

Ramus further sharpened the analogy between himself and Plato by considering the situation of the University of Paris in terms of Plato's Delian episode. While Ramus did not wish the university to suffer from a plague, nor to be delivered by the priestess of Apollo, he nevertheless anticipated the restoration of mathematics and the expulsion of vain, sophistic philosophy. His denial of any presumption on his part – "I am Peter Ramus of Vermandois, not Plato of Athens"[31] – ironically indicated to the reader that one might reasonably confuse the two men. While he

---

[28] Ramus (1569, p. 13): "proindeque e vestigio volare Platonis literae ad omnes familiares in Italiam, in Aegyptum, in Graeciam universam, omnesque praestantes Geometras excitare ad hoc problema demonstrandum."

[29] Ramus portrayed Plato in a similar way in his second *actio* against Charpentier, written the previous year. Telling the story of Delos in order to underline Plato's preeminence in Greek mathematics, Ramus concluded by saying, "And so Plato's letters on this subject flew not only to Italy and the whole of Egypt, but through the whole Greek world, in order to inflame all people to the study of mathematics." (Ramus (1599, p. 419): "Itaque Platonis epistolae ex hoc argumento volare non in Italiam solum et Aegyptum, sed in universam Graeciam, ad omnes mortales Mathematicis studiis inflammandum.")

[30] See, for instance, his letter to John Dee, in Ramus (1599, pp. 174–175); on p. 14 of the *Scholae mathematicae*, he related the information he obtained from their correspondence (that there were no university professors of mathematics in England) and urged Queen Elizabeth to appoint Dee to a royally-funded chair. On p. 66 he recalled his correspondence with Joachim Rheticus over the possibility of an "astronomy without hypotheses" – a project that was so close to his heart that he offered his own chair in the Collège to any astronomer who succeeded in it (*ibid.*, pp. 49–50). The letter to Rheticus may be found in Ramus (1599, pp. 213–218). On Ramus's search for a non-hypothetical astronomy, see Jardine (2001); and Grafton (1997, pp. 261–262), which sets out the historical basis for Ramus's critique of Ptolemy.

[31] Ramus (1569, p. 13): "P. Ramus Veromanduus sum, non Atheniensis Plato." He would make a parallel statement later in the *Scholae mathematicae* (at p. 110), injecting himself into the narrative: "I am Peter Ramus, regius professor at Paris, anxious about the future of the teaching of

could not restore mathematics by interpreting a Pythian oracle, he could (like Plato) call upon scholars – both his fellow countrymen (*domesticos*) and those linked to him by friendship (*familiares*) – to assist him in recalling Plato and Pythagoras to the Academy.

A little later, Ramus returned to the Delian anecdote, much to the same end. Having called upon Queen Elizabeth of England to establish regius professorships of mathematics at the two universities, he reminded his readers that "Plato set the whole world on fire for the study of mathematics by means of that Delian problem of doubling the cube,"[32] before continuing his history of mathematics with the students and followers of Plato. Plato's influence, he seemed to say, was not to be measured so much by his mathematical discoveries, but by the activity he provoked through his patronage and persuasion. Once again, Ramus was drawing an obvious parallel with his sense of his own mission and gifts.

## The Critique of Plato

Ever since the 1543 publications, Ramus had celebrated Plato as his intellectual forebear and model. He purported to recognize in Plato's dichotomous dialectic his own single "method" though he rejected any kind of idealistic interpretation of Plato's universal Forms. In the mathematical prefaces, he reinterpreted Plato's notion of reminiscence in historical terms, to support his own model of gradual change and progress in mathematics, and he lauded Plato as one of the great mathematical authorities. It may seem that in the *Prooemium mathematicum* Ramus was further expanding his positive assessment of Plato, now by depicting him as a member of an international republic of letters responsible for a great reformation in mathematics. In fact, however, there was a major shift of emphasis in the *Prooemium*. Plato occupied a far more ambivalent position. Ramus still considered him a mathematical authority and a model for his own approach to doing mathematics; but now Ramus blamed Plato the *philosopher* for the subsequent decline in mathematics. In the mathematical prefaces, Ramus had laid the blame for the degeneration of mathematics at the feet of Euclid and Theon; in the *Prooemium*, these two "elementators," while still bearing some responsibility for the poor state of mathematics, were cast as mere by-products of Plato's ruinous influence.

Plato, as the heir to three divergent mathematical traditions, had the opportunity to restore and reestablish the primitive state of mathematical knowledge. Yet this opportunity was squandered through what Ramus called Plato's "almost womanly jealousy."[33] Having in his possession a mathematics of unequalled power, Plato

---

mathematics." The context here was again Ramus's efforts to build an international coalition of mathematicians.

[32] *Ibid.*, p. 15: "Plato mundum universum mathematicae studio per deliacum illud duplicandi cubi problema incendit atque inflammavit."

[33] Ramus (1569, p. 18): "ista pene muliebris zelotypia."

fatefully decided that it was to be a subject only for philosophers: "philosophy would be cheapened if mathematics were put into the practical hands of craftsmen."[34] Plato acted just like the ancient Roman *pontifices*, who withheld their calendars from the people or (most significantly, given the confessional context of the *Prooemium*) just like modern theologians, who forbade the laity from studying theology.[35] Wishing to keep mathematics a possession of mathematicians alone, Plato prohibited his students Archytas and Eudoxus from pursuing mechanical solutions to geometrical problems. Mathematics that made use of instruments or approximation looked like the work of artisans or craftsmen; only exact solutions, furnished with proofs, were acceptable to the philosopher – and such "theoretical" approaches were only *accessible* to the philosopher.

Plato's actions cast a pall over the entire subsequent history of mathematics; his "great glory in mathematics was spattered by this foulest of stains."[36] Henceforth, all mathematicians would face a struggle between following natural, practical mathematics, and embracing Plato's elitist mathematics directed towards abstract contemplation. Nowhere is this struggle more apparent than in Ramus's account of the career of Archimedes. Ramus reported all of the remarkable mechanical deeds told of Archimedes, from burning the ships at Syracuse to analyzing the composition of Hieron's crown. He also expressed the highest regard for the ancient mathematician's geometrical works. But even Archimedes could not escape Plato's error, that "the practice of mathematics was not to be shared with the crowd,"[37] with the result that he sought renown for his geometry, not for his mechanics. While he could shake off his Platonic prejudice long enough actually to achieve the most remarkable practical effects, he succumbed to the error in his decision not to *write* anything on the subject – to the enormous loss of posterity.[38]

Plato had quite deliberately opened up a chasm between mathematics and its applications, a schism in mathematics that lasted to Ramus's own day. And this, ultimately, was Ramus's diagnosis for what ailed the modern art. Euclid and Theon's *Elements*, which Ramus in his prefaces blamed for the decline of mathematics, now appeared as only a symptom of Plato's terrible idea. The learned world, cut off from utility, had been left with an ennervated, "speculative" science. Instead of presenting useful results clearly and self-evidently – self-evidence being the very mark of natural, uncorrupted science – mathematicians strove for novelty and cleverness. They departed so far from conformity to natural reason that in order to convince

---

[34] *Ibid.*: "Vilescet philosophia, si mathesis mechanicis opificum manibus exponatur."

[35] *Ibid.*: "Sic pontifices Romani, fastus quondam suos; sic theologi plerique nostri theologiam populo ignotam esse voluerunt." Ramus repeated this charge later (at p. 54), recalling the jealousy of Plato and the "common arrogance of priests, theologians and philosophers" ("ambitionem pontificum, theologorum, philosophorum communem").

[36] *Ibid.*, p. 19: "Maxima igitur Platonis in mathematicis gloria foedissima ejusmodi maculam sibi aspersit."

[37] *Ibid.*, pp. 27–28: "Vetus illa jam inde a Platone mathematicis perversa et praepostera opinio fuit, mathematicae usum non esse vulgo communicandum…"

[38] *Ibid.*, pp. 28–29.

the reader of their truth, they had to write *demonstrations* of their results– casting a pernicious veil cast over simple mathematical truths.[39] The very uselessness and obscurity of this style of mathematics had killed off the last bit of interest in the subject in the schools. Ramus laid the ultimate blame for the continuing neglect of mathematics at the University of Paris at the feet of Plato's "blind ambition."[40] If the abuses in mathematics and the other sciences could be fixed, and if a natural mathematics founded on practice were restored to the universities, then (as Ramus advised Pope Pius V) the rifts in Christendom might also be healed. Rome, filled with mathematicians and other properly trained artists, would attract deeply learned theologians who were also grounded in mathematics and the other liberal arts, while keeping out the sleek and divisive "Aristippuses." These properly trained theologians (grounded, no doubt, in Ramus's methodized theology) would be able to pronounce authoritatively and finally on religious disputes.

> Then Christians will rejoice that heresies and terrible divisions are removed from the Christian religion. Then all will confess that Rome is truly triumphant, and all men will embrace, cherish, kiss this golden pontificate.[41]

## The Heroes of Mathematical Inquiry

Such were the causes of decay; but Ramus also found positive developments in the history of ancient mathematics. Just as he had created a admirable Plato (alongside a harmful, theoretical one), so he recast or even invented the biographies of other ancient mathematicians to suit his polemical purposes.

The heroes in Ramus's history were those scholars whose goals were much like his own: who sought to apply mathematics and extend its utility to ordinary people, or who concerned themselves with its methodical presentation, making it more readily teachable. Among those who devoted themselves to practical mathematics, two stood out: Archimedes (whom Ramus admired with reservations) and the mathematician and engineer Heron of Alexandria. About the latter, Ramus expressed boundless enthusiasm: "this author pleases me especially, because he so effectively and carefully yoked Plato's geometry with Archimedes's mechanics, the art with its use."[42] Such was Ramus's ideal: not simply to pursue mechanics, and still less, of course, to pursue theoretical mathematics for its own sake; but to construct a systematic geometry that was based upon utility and immediately applicable to

---

[39] Ramus's odium for demonstration and his quest for an authentic, undemonstrated mathematics will be considered in fifth and sixth chapters.

[40] Ramus (1569, p. 30): "caeca ambitio."

[41] Ramus (1569, p. 110): "Tum impietates e christiana religione et immanes sectas Christiani sublatas esse laetabuntur. Romam tum denique vere triumphantem omnes confitebuntur ... hunc aureum pontificatum omnes mortales complectentur, fovebunt, osculabuntur."

[42] *Ibid.*, p. 35: "Quamobrem iste mihi imprimis placet author, qui Platonis geometriam cum Archimedis mechanica, qui artem cum usu artis tam solerter atque industrie conjunxerit."

practical ends. Accordingly, Ramus told his readers, he had sought out Heron's works throughout the manuscript libraries of Europe. (In actual fact, most of what Ramus had to say about the works of Heron – his elegantly illustrated work on war machines, the existence of a manuscript of his *Mechanica* in the Vatican Library, another manuscript of Heron's *Geometry* in the hands of Diego Hurtado, imperial legate to Venice – he took directly out of Konrad Gesner's bibliography.)[43] Nevertheless, there are indications in Ramus's paean to Heron that he and his associates had taken a real interest in the ancient mathematician and had studied hitherto little-known works in person. In particular, Ramus said they had edited the Greek text of Heron's *Pneumatica* and fragments of his *Automata*. Based on his limited knowledge of the extant writings, and the rumors of some lost systematic geometrical treatises, Ramus was willing to elevate Heron to a privileged place among the greatest of mathematicians: he was to be ranked not just with Archimedes, but also with Aristotle (who, despite his many failings, founded his physics on mathematical principles) and Leon, composer of the *Elements* used in Plato's Academy.[44]

This last point brings us to the "methodical" teachers of geometry, who loom so large in the *Scholae mathematicae*. Indeed, Ramus's lists of systematizers reflect one of his central concerns in writing mathematical history, namely, to coopt some mathematicians as model teachers while at the same time excluding others. For the basic sequence of Greek mathematicians, he drew especially upon Proclus's list of Greek mathematicians in his *Commentary* on Euclid, though Ramus put these names to very different use.[45] Proclus's text presented the history of mathematics as a series of transmissions from teacher to student, remarkably like the history of philosophy itself as it was imagined in late antiquity. Through this tradition, the sciences had developed progressively from the basely material to the purely spiritual. Ramus's ends were entirely opposed to those of Proclus; yet he plundered Proclus's text to compile a series of "elementators," or composers of "Elements."[46] For many of the mathematicians in the list, Proclus recorded nothing more than their name; and they remain to us little more than names – as they were, of course, to Ramus too. Nevertheless, Ramus loaded them (or, at least, the mathematicians up to the age of Plato) with accolades for collecting and preserving the divine, perfect and undemonstrated mathematics of the first humans. The lists changed slightly throughout the *Scholae*,[47] but Euclid was generally the sixth and penultimate elementator, while Theon took the final spot – and thus much of the blame for the mess that mathematics was in.

The *Elements* would receive sustained abuse in the rest of the *Scholae*, but Ramus's last insult to Euclid was to take away his authorship of them entirely.

---

[43] Gesner (1545), s.v. "Hero Alexandrinus."

[44] Ramus (1569, p. 35).

[45] On the nature of Proclus's list, see p. 1 above.

[46] See, for instance Ramus (1569, pp. 77, 100) for two examples of a list of elementators. The term "elementator" translates Proclus's *stoikheiôtês*.

[47] Most particularly in the inclusion of Pythagoras, which will be discussed in detail below. Compare also the list on p. 35, which includes Geminus as well as Pythagoras.

Ramus thought of the *Elements* as a communally-composed work, as much a product of history as of individual genius. Euclid occupied an insignificant position: certainly not one of the pioneers of the *Elements*, nor yet the architect of its final structure, he was merely a minor transitional figure to whose name the *Elements* had been accidentally attached. Ramus doubted even the authenticity of the minor works attributed to Euclid, "so that nothing is left to him except an empty name."[48]

Given a knowledge of Ramus's past mathematical works, it comes as little surprise to find him elevating the mechanic Heron or denigrating the theoretician Euclid. Even his attack on Plato, though new to the *Prooemium mathematicum*, is hardly out of character; very few ancient authors, if any, escaped Ramus's critical pen. But as Plato's stock sank, another figure gained quite unexpectedly in prominence in the *Prooemium*: Pythagoras. In order to explain this new turn of events, it will be necessary to examine contemporary events at the University of Paris.

## The Institutional Context

Thus far, I have considered the intellectual motivations and concerns underpinning Ramus's history of mathematics and the shape it took in the *Prooemium mathematicum*. Caught up in the religious conflict of 1560s Paris, Ramus looked for a new, universalizing model for the history of mathematics that would parallel his interpretation of the history of the Church. Mathematics, like the knowledge of God, had been given to humanity once and for all at the beginning of time. The historian's task (in both fields) was to discover where fallible humans had obscured the clarity of the message. Plato, at once brilliant and jealous of his intellectual possessions, resembled the medieval theologians whom Ramus deplored in his religious and philosophical works. The teaching of mathematics had to become simple once more. What is more, Ramus had a clear vision of the *kind* of mathematics he wished to promote: an art at once practical and methodical. He clearly shaped his accounts of the *elementators*, of Heron and Archimedes, to support that vision. But alongside these intellectual and religious factors, there were also the difficult academic circumstances in which Ramus found himself in this same period.

On March 11, 1566, the Parlement of Paris decided a bitter dispute between Ramus and his longtime enemy, Jacques Charpentier. At the center of the dispute was the teaching of mathematics, and its place in the Collège Royal, of which Ramus was dean. For Ramus had not been pursuing his mathematical interests in isolation. A long-time member of the Collège, Pascal du Hamel, had held for many years a chair of mathematics; though not particularly brilliant, he, like Ramus, had been a student of the pioneering Parisian mathematician Oronce Fine – who himself had held a chair and the deanship of the Collège when Ramus was first appointed. After

[48] Ramus (1569, p. 39): "ut Euclidi praeter inane nomen nihil admodum relinquatur." In both fifth and sixth chapters, I shall consider why it was that Ramus questioned the significance of Euclid, and how he reimagined the authorship of the *Elements*.

Fine's death, his chair went to Ramus's protégé Jean Pena, who held the position from 1557 until his premature death in 1558; during his brief tenure, he published important editions of Euclid's optical works. After his death, Ramus (while still officially professor of philosophy and eloquence) assumed unofficially much of the teaching of the mathematics professor. Since 1563 Pierre Forcadel, a practical mathematician, had held Pena's chair and had assisted Ramus closely with his mathematical studies – a necessity, we recall, since Ramus required extensive coaching before entering the classroom.[49] Nevertheless, as a student of Oronce Fine, Ramus was the heir of a tradition of reform-minded mathematicians at the University.[50]

In 1565, du Hamel died, to be replaced by a Sicilian named Dampestre Cosel. While he was a competent enough mathematician, he was unable to speak either French or Latin, and his teaching was, understandably, a disaster. Ramus, as dean, agitated successfully for his dismissal, but found that even worse was to come. According to Charpentier, when Cosel found his position to be untenable, he approached the Cardinal of Lorraine (then acting as the king's agent for all matters pertaining to the Collège), tendered his resignation and recommended Charpentier as his successor; the Cardinal quickly acted upon this suggestion.[51] By Ramus's more jaded account, Charpentier bought the position from Cosel.[52] Whatever the case may have been (and Ramus's version, though outrageous, seems to have been closer to the truth), Ramus found that his colleague in his most cherished subject was one of his bitterest opponents.

Charpentier was a protégé of Pierre Galland, one of the principal engineers of Ramus's early ban from teaching philosophy. When Ramus was elevated to the Collège (thereby thwarting the schemes of the Galland party), Charpentier and Galland attacked his students, denying them the right to graduate from the university with a degree. In response, Ramus had cheerfully taunted Galland and Charpentier in a public oration, referring to Charpentier only as Galland's "beardless acolyte," with obvious implications of pederasty.[53] This set the unedifying tone for his quarrel with Charpentier. Over the years the two men clashed many times, their arguments often spilling over into pamphlet wars involving their supporters and students as well.

Their differences went far beyond personal animosity. Charpentier was a medical doctor and an Aristotelian philosopher, who, in his own studies, had concerned

---

[49] See p. 20 above.

[50] On the professors of mathematics at the Collège Royal, see Pantin (2004), especially the table of holders of the chairs on p. 200. On Fine and reform, see Margolin (1976).

[51] Charpentier (1566, sig. D4v).

[52] Ramus, *La Remonstrance faite au conseil privé en la chambre du Roy, au Louvre le 18 janvier 1567*, Paris (1567, pp. 14–15); cited by Skalnik (2002, p. 83).

[53] In his 1551 speech *Pro philosopica disciplina*, Ramus (1599, pp. 255–323). On Charpentier's association with Galland and his animus against Ramus, see Waddington (1855, p. 41; and pp. 73–75) on the circumstances of the speech itself. Charpentier's obsessive hatred for Ramus was attested even by a sympathetic biographer like Masson (1638, pp. 272–274), who judges that Charpentier's greatest vice was his implacable anger, which he himself admitted could not be assuaged by any philosophical remedies.

himself with the reconciliation of Aristotle and Plato.[54] His intellectual opposition to Ramus was founded on Ramus's claim to have shown that there was one single method common to Plato and Aristotle; Charpentier, dedicated though he was to the unity of the philosophers, did not agree and was well enough read in the authors to mount a formidable challenge. He remained thoroughly in the "scholastic" camp of the University, and taught entirely through the reading and commentary of Latin translations of philosophical texts; his own publications, likewise, paid no attention to the authentic texts of the philosophers. He also remained unmoved by the mathematical side of Platonism; in his natural philosophy he was an Aristotelian through and through. In one of the pamphlets issued in an earlier dispute with Ramus, Charpentier went so far as to proclaim proudly (though, as it turned out, rashly) that he was "*analphabêtos*" and "*ageometrêtos*," that is, illiterate in Greek and ignorant of geometry,[55] intending by these Greek words to differentiate himself from Ramus and his followers. From Ramus's point of view, even on purely intellectual grounds Charpentier was a disastrous choice for a chair of mathematics in the Collège.

Ramus relied on the mathematical expertise of those around him, both to help him prepare classes for his students and to work with him on mathematical publications. This was a matter of public knowledge; and Charpentier, not surprisingly, took delight in taunting Ramus for his lack of mathematical expertise and his dependence on others.[56] But Ramus had long maintained (as we have seen) that progress in mathematics was a slow, communal effort, and he freely admitted to the assistance he had received from his various colleagues.[57] Indeed, it seems that Ramus, in these latter years, saw himself and the two mathematical professors as a unit within the Collège, collectively pursuing the mathematical reform of the arts that he had foreshadowed in his various publications. By 1566 Ramus himself, although not a professor of mathematics, was teaching elementary classes in the rudiments of geometry and arithmetic, followed by readings of Aristotle's *Mechanics* and Archimedes's *Sphere and Cylinder*; and he had told his allies that he intended in the future to continue on to cover optics as well.[58] However slender Ramus's own mathematical talents may have been, his informal mathematical society within the Collège was achieving remarkable results. When Charpentier seized one of these chairs (without even a

---

[54] Ong (1958a, p. 220).

[55] In Charpentier (1564, fols 3v and 11r–v).

[56] For instance, in one of his orations he wrote that if Ramus demanded wide mathematical knowledge from a professor, then he should not be a teacher himself "since not only is he not well established in this subject through long and assiduous practice, but (as his teachers will attest) he can hardly even parrot faithfully what has been dictated to him at home." (Charpentier (1566, sig. G2r): "Quoniam hic non modo in ea non est longo usu et assiduo confirmatus, sed vix adhuc potest, quod eius magistri testantur, domi dictata, suis fideliter recitare.")

[57] See particularly his *Actio secunda* against Charpentier, where he "admits, or rather proclaims" that he has received help from others (Ramus 1599, p. 431); and his preface to the 1569 *Arithmetica*, acknowledging the assistance of Pena, Forcadel and Risner (*ibid.*, pp. 135–136).

[58] This is according to the testimony of Charpentier, in Charpentier (1566, sig. E4v). Charpentier is not, of course, an objective source on Ramus's teaching. But he can be trusted here since he is admitting, almost despite himself, that Ramus has adopted an ambitious mathematical curriculum. (In a final, catty remark, he laments that Ramus had made such an effort for so few students).

pretense of interest in teaching mathematics) it seemed that Ramus's *de facto* control over mathematics within the Collège and his program of reform for the University had been thwarted.

In order to block Charpentier's tenure of the chair, on March 8 Ramus obtained an injunction from the king stating that professors of the Collège had the right to examine all those who wished to join their ranks, and to reject those who failed to meet their standards. This right of examination would become something of an obsession for Ramus: the principal remedy, as he saw it, against academic abuses in the Collège. In order to impose this requirement on Charpentier, Ramus brought him before the court of the Paris Parlement on March 9, where he delivered his first, brief *Actio mathematica*, to which Charpentier replied with the first of his three *orationes*.[59] On the eleventh, they reconvened, Charpentier opening with his second *oratio*. At Ramus's insistence, Parlement then went into public session for the rest of the day, and he delivered his long, second *actio*, in which, before the people of Paris, he praised their native mathematical ingenuity and lamented its betrayal by the university.[60] Charpentier replied with his third and final *oratio*. After deliberation, the Parlement found in favor of Charpentier, by issuing an *arrêt* confirming him as professor of mathematics.[61]

The Parlement made some concessions to Ramus's demands. Throughout the case, Charpentier had argued that the subject matter of each chair was not fixed, so that there could be no objection to him teaching Aristotelian metaphysics instead of mathematics, as he intended.[62] On this question, Parlement sided with Ramus, saying that there should indeed be two permanent chairs of mathematics within the Collège, neither of which could be taken over for other disciplines. Moreover, they even agreed that prospective professors of the Collège should be required to submit to an examination, confirming the decree Ramus had obtained from the king. Yet this was hardly a victory for Ramus, at least so far as Charpentier's tenure was concerned; Parlement had conceded the general principles only to subvert their particular application. For although the *arrêt* stipulated that Charpentier must teach mathematics while holding this chair (despite the fact, as they admitted, that he was woefully ill-prepared to do so and in fact intended to teach philosophy instead), nevertheless, Charpentier need only lecture on *something* mathematical within the first three months of his tenure. This should not pose any difficulty to him (the *arrêt* went on) since he was an intelligent man, and, unlike eloquence, mathematics required no great skill, just the ability to draw with a pencil! Lastly, while new professors in *general* should submit to an examination, "for many very sound reasons and valid considerations" Charpentier was to be exempted from this requirement.

---

[59] The chronology of the case is complicated by the fact that Ramus made several errors in dating his orations according to the Roman style. The order of the speeches as I give them here is based on internal cross-references in the speeches.

[60] Ramus (1599, pp. 420–422). See n. 16 above.

[61] The text of the *arrêt* is in Waddington (1855, pp. 176–178). See also Pantin (2004, pp. 193, 202).

[62] See particularly the second oration: Charpentier (1566, sig. D4v). He noted that even Ramus, Professor of "Philosophy and Eloquence" had taken the chair of a Professor of Hebrew.

This was not the end of the matter, as Charpentier did not fulfil even the lax stipulations of the *arrêt*. According to Ramus, he had originally promised that he would lecture on Aristotle's *De caelo*, Proclus's *Sphere*, Euclid's *Elements* and Sacrobosco. The undertaking he read into the *arrêt* said, however, that he would lecture on Aristotle and Proclus, *or* Euclid and Sacrobosco. Near the very end of the three-month period, Charpentier began some lectures on *De caelo* and the elementary *Sphere* of pseudo-Proclus, thus satisfying the letter of the *arrêt* , he claimed. Ramus was outraged; but the President of Parlement, Christophe de Thou, persuaded him not to bring another suit before the court.[63] Instead, he responded, in January 1567, by appealing directly to the Privy Council, and publishing the text of his suit.[64]

This was the last attempt at legal redress; henceforth both men fought their corners entirely through the printing press. Ramus, of course, was now the only one with a grievance and substantive complaints to air; Charpentier's responses devolved into little more than gratuitous personal abuse.[65] These published responses were not restricted to mere pamphlets; Ramus's *Prooemium mathematicum*, published a year after the trial, was part of his response to Charpentier; his preface to the work, asking for direct royal intervention in the teaching of mathematics at Paris, implicitly connected his exhortation to mathematical studies with his defeat in court.

The notion of appointing Charpentier to a chair of mathematics – and maintaining him there, despite his cavalier disregard of the terms of the chair – seems absurd, even capricious. Charpentier was, without a doubt, entirely unqualified to teach mathematics, just as Ramus so colorfully asserted in his many controversial pamphlets and orations related to the succession to the chair; moreover, as Charpentier himself admitted, he had obtained the chair without even the *intention* of teaching mathematics.

Modern scholarship has tended to take Ramus's side of the issue. Charles Waddington, his nineteenth-century biographer, saw Charpentier's actions as base and dishonorable, deliberately harming science out of a "cynisme révoltant," while Ramus, on the other hand, was motivated above all by his concern for academic honesty and rigor.[66] Waddington was never less than admiring of Ramus, so that his support for his subject's position is quite predictable. Walter Ong, by contrast, was generally dismissive of Ramus's mathematics. In one of the very few references to this aspect of Ramus's studies that he made in his intellectual biography of Ramus, Ong marvelled at his "mysterious" growing interest in mathematics from the early 1560s, "mysterious because he was so ill-educated in mathematics."[67] And elsewhere he wrote that Ramus's accusation of incompetence against Charpentier and Cosel in the *Actiones duae* "is interesting in view of the fact that incompetence was

63 Waddington (1855, pp. 178–179); Girot (1998, pp. 70–71).

64 Ramus, *La Remonstrance*, extracts from which are edited in Waddington (1855, pp. 411–417).

65 Waddington (1855, pp. 178–181). The only complete and accurate chronology of the case and the subsequent pamphlet war is found in Girot (1998).

66 Waddington (1855, p. 181).

67 Ong (1958a, p. 27).

one of the grounds on which Ramus himself, 22 years before, had been suspended from teaching philosophy,"[68] thereby insinuating that the entire case was nothing more than a tit-for-tat act of revenge on Ramus's part. Yet even Ong had to admit that, so far as the substantive issue between him and Charpentier was concerned, Ramus was in the right. Charpentier knew "even less mathematics than Ramus," who at least had managed to inspire others to achieve what "he himself could not realize."

For both Ong and Waddington, the case turned upon the ability to teach mathematics. Both scholars agreed on the justice of Ramus's cause: mathematics is a fine thing, and Charpentier knew nothing about it. More recent scholarship has concentrated on the larger issues that lay behind the debate. In a recent study, Skalnik has argued that Ramus's central ideological commitment was to a notion of "merit" (which Skalnik associates with the court of François I) against a retrenching of aristocratic privilege. He argues that the motivations of both actors in the tussle over the chair can be illuminated by these opposing political or social ideologies.[69] Ramus's attack on Charpentier was thus prompted not so much by the need to safeguard mathematical teaching at the University, as by the irregular means by which the chair was obtained: Charpentier had done nothing to *deserve* it. Charpentier, on the other hand, who reminded Parlement *ad nauseam* of his powerful patrons, stood for unashamed privilege. Skalnik's sympathies lie with Ramus: Charpentier had obtained the chair through a private transaction and without "consideration of qualifications,"[70] and the outcome was a foregone conclusion anyway, since this confrontation between a François I meritocrat and the "elite oligarchy of the Old Régime" was decided by the "venal Parlement of Paris."[71]

Girot, in his very careful study of the dispute, is the only modern scholar to conclude that, in fact, Charpentier's arguments were stronger than Ramus's, given the context in which they were made. That is, Charpentier judged his defense perfectly with respect to the political situation of 1560s Paris, and the verdict was the only one possible after all the arguments had been heard. According to Girot, the argument was really about *authority*. Charpentier insisted that the king (or his agent, the Cardinal of Lorraine) could appoint anyone he wished to the Collège, to the position of a *royal* professor. The choice of the king was absolute; while he might listen to advice, he could in no way be compelled to take it. So too could he judge a candidate's qualifications according to any standard he pleased. This was how Charpentier himself viewed the matter: by raising opposition to Charpentier's appointment, Ramus was only revealing his own anti-authoritarian bent.[72]

According to Charpentier, just as Ramus delighted in subverting the authority of Aristotle and the ancients, so too was he now taking advantage of this situation to undermine the power of the king, substituting his own whim and the malleable

---

[68] Ong (1958b, p. 357).

[69] Skalnik (2002, pp. 81–87).

[70] *Ibid.*, p. 83.

[71] *Ibid.*, p. 87.

[72] See, for instance, Charpentier (1566, sigs B2r, C2r–v, H3r–v).

opinion of a committee of professors.[73] Moreover, Charpentier argued that he had already been "put to the test" by his long, successful career in the University. In trying to set up his own examination, Ramus was rejecting another source of established authority – the University – in favor of his own opinion. Finally, Girot shows that on the substantive matter of the institutional history of the chairs, Charpentier was correct, not Ramus. There was no clear identification of the subjects to be taught by each professor; thus there was nothing irregular in Charpentier not teaching mathematics, even though he succeeded to a chair formerly held by a mathematician.[74] In summary, though Ramus wanted to present himself as an intellectual reformer, facing a stubborn sophist, Charpentier reframed the case as a confrontation between a loyal subject and a political partisan, or (as was constantly implied) a good Catholic and a disloyal Protestant.[75] However specious his reasoning may have been, Charpentier brilliantly recast the terms of the debate. Presented with a Protestant Ramus who sought to limit the sovereignty of the king in favor of that of an assembly, Parlement could not but award the case to Charpentier.

By removing the debate from the realm of the history of ideas or of science, to that of political history, Girot provides an entirely satisfying account, in which the motivations of all the actors are explained – and in which the *arrêt* of Parlement, granting Charpentier the chair even while recognizing his mathematical deficiencies, does not seem entirely perverse. Skalnik's social considerations are equally valuable, and I shall argue that the themes of legitimate authority and qualification to possess it are central issues in the debate, especially in Ramus's *Prooemium mathematicum*. It must be observed, however, that both these recent treatments omit entirely any consideration of the intellectual substance of the debate. In fact, insofar as it was about *anything* substantive, the debate focused on the history (or rather, the imagined *prehistory*) of mathematics. In this at times fictive historical narrative, the figure of Pythagoras emerged as a key – and hotly contested – protagonist; and so it is to this figure that we must now turn.

## Claiming Pythagoras

The appearance of Pythagoras in the Ramus-Charpentier debate of 1566, and the intense interest Ramus expressed in his career in the 1567 *Prooemium mathematicum*, are both somewhat surprising. In his early accounts of the history of

[73] Girot (1998, pp. 73–74). Note, for instance, in Charpentier's first oration before Parlement (Charpentier, 1566, sig. B2r) that he equated Ramus's insistence on holding an examination with a desire to usurp regal powers for himself. In the third oration (*ibid.*, sig. D4r) he compared Ramus's tenure as dean to the madman who, just the other day, had gone running through the streets proclaiming himself king of France. Through this comparison he associated Ramus, as usual, with unrestrained passions and delusions of grandeur, but also with treasonous ambitions.

[74] Girot (1998, pp. 79–81). Skalnik also acknowledges that Charpentier was correct on this and other points of institutional history and practice. See Skalnik (2002, n. 57 on pp. 85–86).

[75] Girot (1998, p. 74).

mathematics (surveyed in the previous chapter), Ramus said hardly a word about Pythagoras. In the preface to his 1544 edition of Euclid's *Elements* (one of his first, brief attempts at a narrative of the origins of the art), Pythagoras's name appeared only once, in a list of mathematicians who flourished in Greece, long after the art originated among the ancient Hebrew patriarchs. Ramus presented an almost identical list some 10 years later, in the preface to his *Arithmetic* of 1555.[76] In these writings, the traditional figure of Pythagoras, mystic and numerologist, held very little attraction.[77]

Even a decade later, now thoroughly engaged in the teaching of mathematics in Paris and embroiled in dispute with Charpentier, Ramus still evinced almost no interest in Pythagoras. In his second *actio* against Charpentier (1566), Ramus provided the Paris Parlement with a brief history of mathematics from the patriarchal age. Pythagoras and his contemporary Thales figured only incidentally, their role as links between Egypt and Greece described in a single sentence. Ramus focused on another traveller to Egypt, Plato, whom he identified as the key figure responsible for the foundation of mathematics in Greece, both through his own efforts and through his support of the mathematicians in his Academy.[78]

In the 1567 *Prooemium mathematicum*, however, Ramus paid far more attention to Pythagoras (even as he cast doubt on the wholesomeness of Plato's influence). One has to read this book as a fruit of defeat. It stands out above the scurrility of the Ramus-Charpentier pamphlet war, but it cannot be separated from the case. The *arrêt* awarded Charpentier his chair, even acknowledging his lack of mathematical expertise – a commodity which the Parlement set at a very low value. Ramus's arguments on the continuity of the mathematical chair had carried no weight; and his impassioned plea to the court to recognize the centrality of mathematics in history and in the business of life had been ignored. The *Prooemium*, addressed to the Queen Mother and pleading for direct royal intervention, was at once a defense of mathematics and an attack on his opponents. It carried on the dispute, using the weapons of historiography and textual criticism. Here, Pythagoras became a crucial protagonist in Ramus's history of mathematics. No longer merely a figure of transition (and even in that role rather overshadowed by Plato), Pythagoras came to stand for Ramus's ideal teacher of mathematics, his ancient school held up as a model for the University of Paris. This new prominence requires some explanation, not least because Pythagoras, seen in the Renaissance as an austere religious

[76] First mathematical preface, Ramus (1599, pp. 120–121): "Hinc tot, tamque excellentia ingenia excitari, Thaletis, Pythagorae, Hippocratis, Platonis, Eudoxi, Ptolemaei, Euclidis, Archimedis, aliorumque innumerabilium coeperunt." Second mathematical preface, *ibid.*, p. 121: "haec tandem Graecorum et Italorum, Thaletis, Pythagorae, Anaxagorae, Hippocratis, Platonis, Archytae, Aristotelis, Euclidis, Philolai, Archimedis, reliquorum omnium (de quibus Proclus scripsit) celebrata gymnasia fuerunt."

[77] In his early neglect of Pythagoras, Ramus was following the lead, it seems, of Regiomontanus who, in his 1464 oration on the history of mathematics, passed over Pythagoras in a single sentence (as noted in first chapter above).

[78] Ramus (1599, pp. 419–420).

prophet or an unworldly number mystic,[79] hardly seems the kind of figure that would interest Ramus, whose concerns were so oriented towards the practical application of mathematics and other arts.

Ramus's new interest was sparked by a polemical line of argument that Charpentier had introduced in the course of their dispute. Charpentier's unguarded, printed admission that he was "ungeometrical" and "unlettered" had haunted him through the hearing of the case and the ensuing pamphlet war. In his first oration he rather feebly batted the offending words away: if he himself was illiterate, then Ramus was even worse.[80] In the third, he met them head on, giving his apparently ill-chosen words a novel twist: they were, he claimed, meant ironically. In Ramus, he explained, he had found a critic who pretended to universal knowledge, and in particular, expertise in mathematics. Ramus's claims did not, however, match up with reality. In fact,

> I knew that he had often been struck dumb while at the lectern, that along the way he would mislay the very things he had just learnt from his teachers; that on countless occasions he had been forced to botch his way through a mathematical demonstration, because he hadn't practised it enough; and that often in his lectures he would completely contradict something that he had affirmed with, it seemed, great confidence in an earlier lecture.[81]

Ramus was, Charpentier concluded, just like the sophists who challenged Socrates, or like the so-called "wise" with whom Pythagoras had disputed. Both Socrates and Pythagoras were, of course, highly learned in the very disciplines in which their detractors pretended expertise. Yet they disarmed these false claimants to wisdom by adopting an ironic pose: Pythagoras said he was not wise, but a "lover of wisdom," while Socrates professed to know only that he knew nothing. In neither case was this literally true: both men had positive, substantial knowledge which they passed on to their schools, and which was still studied. Charpentier claimed he had meant to use the same ploy of false modesty with Ramus. Of course (he now insisted) he was not truly illiterate and ungeometrical. But, like Pythagoras and Socrates, he had faced an opponent both entirely ignorant of the arts he professed, and absurdly confident in his skill at professing them – a situation in which irony was the only

[79] See Heninger (1974); Riedweg (2005, especially pp. 129–132); and Joost-Gaugier (2006, especially pp. 66–76). The figure of Pythagoras was quite malleable; not long before Ramus, Johannes Reuchlin had claimed he had brought Pythagoras back to life in his presentation of Kabbalistic wisdom, insisting that "Kabbalah and Pythagoreanism are of the same stuff" (Jones, 1983, p. 19).

[80] Charpentier (1566, sig. B4r).

[81] Charpentier (1566, sig. G3v): "Certo sciebam hunc in Cathedra Mathematica saepe obmutuisse, quod in via de manibus excidissent ea quae a magistris paulo ante acceperat; millies etiam inter docendum coactum fateri, Mathematicam descriptionem parum feliciter succedere, quod in hac non esset satis exercitatus; nec minus frequenter posteriore lectione ea omnino invertisse, quae superiore magna animi confidentia videbantur esse constituta." In the same vein, in the aftermath of the case one of Charpentier's anonymous supporters recorded how Ramus lost the thread of a geometrical proof in front of his class, and, entirely out of resources, stood agape and "dumber than a fish" in front of his bemused students. Anonymous (1567, p. 9): "... dum videlicet susceptae propositionis demonstrationem nulla ratione potuisti exponere, sed pisce mutior factus, illico de cathedra descendisti."

possible response. Any other reply would have meant engaging with Ramus as if they were on the same intellectual level.[82] Charpentier's self-identification with Pythagoras – and, even more, with Ramus's life-long role model Socrates – must have been galling to Ramus. Beyond merely this, Charpentier had demoted Ramus to the role of a historical nobody, an anonymous sophist or pretender to wisdom. This was the catalyst for Ramus at last to pay attention to Pythagoras, to add him to his small pantheon of ancients worthy to be emulated, and to recast him as a prototype not for Charpentier, but for his own career and scholarly agenda.

## Pythagoras the Ramist Schoolmaster

Throughout his career, Ramus had insisted that mathematics should be an integral part of a liberal arts education. But the debate with Charpentier raised the stakes considerably. The *arrêt* of the Paris Parlement had declared that mathematics was *not* an art of the same difficulty as rhetoric, and required only the skill of drawing with a pencil; hence, it concluded, Charpentier (or just about anybody else with a modicum of intelligence) could teach mathematics without any special training whatsoever. In the *Prooemium mathematicum*, his extended response to the events of 1566, Ramus insisted that mathematics was not just an art, but the foundation of *all* the arts. In assembling arguments to support his contention, Ramus looked to Proclus's *Commentary*, where a single remark about Pythagoras attracted his attention. Proclus had written that Pythagoras was the first to make mathematics a liberal art. Ramus expanded on this, saying that Pythagoras obtained renown not just for his discoveries in geometry and arithmetic, but also

> because he first brought the mathematical philosophy into the form of a liberal art, and opened a school in which young people might receive a training both honorable and noble.[83]

Proclus had not elaborated on what he meant by a "liberal art" (*doctrina liberalis*), or how exactly Pythagoras had made mathematics such an art. But the meaning was sufficiently clear to Ramus, for whom the term was all but synonymous with an art taught in a school. Indeed, his repeated criticism of the other liberal arts had been that they did not observe the disciplinary boundaries and order of presentation that he insisted upon, and hence were *unteachable*. Proclus's cryptic statement could make sense to Ramus only if Pythagoras had opened a school, where he had been the first teacher of mathematics to young men.

If Pythagoras had opened a school, the question arose (particularly for Ramus) of the kind of teaching and learning that went on there. Proclus said nothing about the nature of Pythagoras's teaching, but many other ancient authors had indeed written

---

[82] *Ibid.*, sigs G3r–v.

[83] Ramus, *Scholae mathematicae*, p. 7: "... quod mathematicam philosophiam in speciem liberalis et ingenuae doctrinae primus redegerit, ludumque aperuerit, in quo juventus tam honestas, tamque nobiles exercitationes haberet." See Proclus (1992, pp. 52–53).

on this very subject (without, of course, imagining that Pythagoras was running a school in quite the same way that Ramus imagined he did). Ramus turned to the *Attic Nights* of Aulus Gellius, where he found a very peculiar description of Pythagoras's system for selecting students worthy of admission to his school: he examined the shapes of their faces and the disposition of their bodies, applying physiognomic principles to determine their character and suitability as students. That is as much as Gellius tells us about this practice, and one might almost expect Ramus to omit this rather strange detail. In fact, however, he expanded upon it, combining it with another anecdote taken from the same source to conclude that Pythagoras's intention in examining his applicants in this way was "to ensure that unrefined, unperceptive and ungeometrical men (*ageômetrêtoi*) should not abuse the leisure and learning dedicated to so liberal an art."[84]

No one who had been following the debate between Ramus and Charpentier could miss the fact that *ageômetrêtos* was precisely the term that Charpentier had rashly used to refer to himself. In Ramus's version of the Pythagoras legend, as he was developing it in these pages, Charpentier, far from being another Pythagoras as he had claimed, resembled instead the students that the ancient mathematician had refused to teach. The physiognomic entrance exam that Ramus described so positively parallels the examination he wished to impose on all candidates for the Collège Royal, by which he had hoped specifically to exclude Charpentier. Ramus reiterated the connection between Pythagoras's school and contemporary Paris a few pages later, when he recalled the famous sign over the door of Plato's Academy, forbidding entry to those without geometry. Plato, said Ramus, was emulating Pythagoras in keeping out the *amousoi, atheôrêtoi* and *ageômetrêtoi*; yet the University of Paris (he regretted) made no efforts to keep such men out.[85]

As well as standing in for a more discerning Collège and University, Pythagoras's school provided a model for teaching at the University of Paris. From Diogenes Laërtius, Ramus learned that Pythagoras wrote three classes of treatises: *paideutikon, phusikon* and *politikon*.[86] Ramus was only concerned with the first, in which he imagined (absent any actual information about it) that Pythagoras set out the pedagogical principles of his school. More specifically, the "form of liberal learning" contained within the *paideutikon* was the division of the school into distinct ranks. After passing a period of silence, Ramus explained, the students abandoned their initial title, *akoustikoi*, and took on that of *mathêmatikoi*, from the knowledge of mathematics they had acquired during their years of silent study. Then, once they

---

[84] *Ibid.*, p. 7: "Non quosvis ait Gellius libro primo capite nono in disciplinam admittebat, sed *ephuseognômonei* ex oris et vultus ingenio ... ne *amousoi, atheôrêtoi, ageômetrêtoi* otio et ludo disciplinae tam liberalis abuterentur." In Gellius, the anecdote about physiognomy occurs at the beginning of *Noctes atticae* I. 9; at the end of this chapter on the Pythagoreans, Gellius records a saying of his friend Taurus, that modern philosophers were *amousoi, atheôrêtoi* and *ageômetrêtoi* in comparison with the followers of Pythagoras.

[85] *Scholae mathematicae*, p. 12.

[86] See *Vitae*, VIII. 6.

had mastered physical studies, they were called *phusikoi*. Finally, they studied the ruling of cities and states, and were then called *politikoi*.

Ramus claimed to be basing his account of the arrangement of the school on Gellius, but he made one significant change to the ancient account. Gellius had stated that the students received their titles when they *began* to engage in a particular activity, so that the *mathêmatikoi* (for instance) were so called once they started to study mathematics, and so long as that was their primary occupation.[87] Ramus, on the other hand, wrote that they received their titles on *completion* of those studies. According to his account, the students completed a mathematical education while still *akoustikoi*. Then, as *mathêmatikoi*, they studied the natural world; when those studies were finished, they became *phusikoi*. As Ramus reinterpreted them, the titles marked off discrete units within the Pythagorean curriculum: mathematics was something to be mastered before moving on to the next subject on the syllabus.

In Ramus's carefully contrived account, Pythagoras's pedagogical instincts (as supposedly recorded in his lost *paideutikon*) conformed precisely with his own. Ramus required just such a strict division of curriculum subjects by his second "law of method," the "law of justice" or homogeneity (in Greek, *kath'auto*). It was by application of this law that he could lambaste Aristotle on almost every page of his *Scholae in liberales artes* for including logical material in the *Physics*, or theology in the *Metaphysics*. He made the contemporary relevance of Pythagoras's curriculum explicit, writing

> If only that *paideutikon* of his, the foundation of a liberal institution, had been a little more carefully observed; then our schools would not have lacked the true elements of humane learning for so long.[88]

Ramus trumpeted the virtues of Pythagoras's school both because it took the liberal art of mathematics as the foundation of all learning, and because it supposedly imposed a rigid, Ramist distinction between disciplines. And there was yet another way in which he imagined that Pythagoras had been the perfect Ramist professor. Immediately following the passage just quoted, Ramus explained that, in Pythagoras's day, there were no studies of grammar, rhetoric or dialectic. Instead, the *initia* and *elementa* of learning were in mathematics; and the completion was in physics. (Politics was an extra subject that could be studied after the principal studies, he explained). In the modern university, Ramus accepted, the elements needed to be learnt from grammar, rhetoric and dialectic.[89] But still, he argued,

---

[87] Gellius, *Noctes atticae* I. 9: "Hi dicebantur in eo tempore *mathêmatikoi*, ab his scilicet artibus quas iam discere atque meditari inceptaverant."

[88] *Scholae mathematicae*, p. 8: "Cuius utinam *paideutikon* illud liberalis et ingenuae institutionis fundamentum, paulo diligentius ab hominibus attenderetur, propria humanitatis elementa tandiu a scholis nostris nequaquam abessent."

[89] In his Ramus (1559, fols 44v–45r), he claimed that the ancient Gauls taught the liberal arts in their native language; if they had written down their teachings, it would be possible for the French to learn the arts in the vernacular, without the years now needed for the study of Latin and Greek grammar.

there was no excuse for omitting mathematics and going straight to physics and politics. Mathematics was the *elementa et fundamenta* of physics and politics, and Pythagoras did not think anyone could become a physicist or politician without first mastering mathematics. It was a scandal that graduates of the University of Paris could be called "masters" when Pythagoras would not recognize them as educated even in the rudiments of philosophy.

Ramus's ideal curriculum – if the Pythagorean *schola* were transferred to the banks of the Seine – thus consisted of training first in the linguistic arts of the trivium, a concession to necessity; then mathematics, and finally physics, each taught in discrete, consecutive units. After that, students could pursue other subjects like politics. But Ramus did not discover this admirable curriculum for the first time in Aulus Gellius. In fact, this was *precisely* the curriculum he had advocated in his oration *Pro philosophica disciplina* of 1551, marking his return to philosophical teaching and writing (and also written, as it happens, in response to continuing obstruction from Charpentier and others).

In that oration, Ramus outlined a 7 year course of study, divided into distinct stages, which he proposed as a model of reform for the University of Paris.[90] Students would move from the study of language to dialectic in the fifth year, mathematics in the sixth, and physics in the seventh. In accordance with Ramus's strictures on "homogeneity," the subjects at each stage were to be kept rigorously apart from each other; there was to be no teaching of rhetoric in a grammar course, for instance, or vice versa.[91] Nevertheless, each successive stage would build on that which had gone before. The use of the three linguistic arts pervaded all the subsequent philosophical study, even if their teaching was to be carefully segregated. Students engaged in learning mathematics would be expected to master geometry, of course, but also would be required to declaim on mathematical subjects (much as Ramus himself would later do in the *Prooemium mathematicum*).[92] And the "physics" studied in the final year would have a much more mathematical flavor than traditional university teaching of natural philosophy. Aristotle's *Physics*, the standard university text, he dismissed as being merely filled with captious logical arguments. Instead, students would extract the natural phenomena from Aristotle's *Meteora*, *De anima* and *Parva naturalia* (suppressing all of Aristotle's irrelevant arguments), and master Euclid's *Optics*, *Catoptrics* and work on musical harmonies, so that a "true physics, founded on mathematical reasoning, will be taught and practiced."[93]

Of course, when he wrote the *Pro philosophica disciplina* in 1551 and described his ideal curriculum, Ramus was not thinking of Pythagoras. Indeed, it seems that at this point he had hardly given any thought to Pythagoras at all. But in the *Prooemium*

---

[90] See n. 53 above for this oration (the text of which is in Ramus 1599, pp. 255–323).

[91] Ramus (1599, p. 170): "Nec in isto rhetorico studio grammaticas regulas permiscemus...;" p. 171: "... et Dialecticae inventionis dispositionisque praecepta, quae Rhetores in rhetoricis artibus parum distincte confuderant, in dialectica arte proprie et perspicue tradimus."

[92] *Ibid.*, p. 177.

[93] *Ibid.*. "... [volumus] Physicam veram, mathematicis rationibus fundatam doceri et exerceri."

*mathematicum*, with a little bending of the historical evidence, Ramus was able to claim Pythagoras as the originator of the very reforms he wished to institute in the University, molding Pythagoras's school until it looked like the ideal university of his 1551 oration. Parrying Charpentier's presumptuous self-identification with Pythagoras, Ramus showed that Pythagoras actually prefigured Ramus, not Charpentier, the "ungeometrical" student excluded from a true liberal education. In Pythagoras, Ramus both found a wry rejoinder to Charpentier, and, much more importantly, discovered (or planted) deep historical roots for the Ramist intellectual and educational program.[94]

As well as making Pythagoras the original Ramist schoolmaster, Ramus also cast him as the first "elementator," that is, the first to construct a collection of theorems that looked much like the *Elements*. His intent, once again, was to draw a direct line between the Ramist program and the very beginnings of mathematics. For Ramus had devoted own career as a master of the liberal arts to writing textbooks, or to improving upon those already written by making them clearer and more "methodical." His dialectic, under almost annual revision, grew out of the humanist dialectics of Valla and Agricola,[95] and his work on mathematics began (as we have seen) with an edition of Euclid's *Elements* and continued to be revised, in accordance with his methodological principles, until the end of his life.[96] In his reconstruction of the proto-history of mathematics, Ramus imagined that Pythagoras, the first teacher of mathematics as a liberal art, occupied himself in much the same way. But Ramus had very little evidence that Pythagoras had written a textbook of mathematics. Proclus stated quite unambiguously that Hippocrates of Chios was the first person to write a collection of *Elements*, and he flourished a generation or more *after* Pythagoras.[97] In order to secure Pythagoras's position as the founder of the tradition of mathematical textbooks, Ramus gently massaged the evidence over the course of the *Prooemium*, making Pythagoras a kind of "proto-elementator" standing at the head of the line of historically-attested authors of mathematical *Elements*.

Ramus's biography of the first attested elementator, Hippocrates of Chios, described his colorful life and surveyed all his mathematical achievements, but singled out his authorship of a book of *Elements* as the most important of all his achievements. It was here that Ramus, while ostensibly describing the magnitude of Hippocrates's accomplishment, reintroduced and reimagined Pythagoras:

> The first teacher of mathematics in a school was Pythagoras, but (as it is only fair to believe about the very beginnings) he was not entirely proficient, so that he is not called an "elementator;" but whatever the case may have been, Hippocrates was not at all put off by the

[94] Two years later, Ramus would use a similar line against the Aristotelian Jakob Schegk, whom he enjoined to keep a modest, "Pythagorean silence" until he had mastered sufficient mathematics to express a worthwhile opinion on philosophy. See Ramus (1599, pp. 205–206).

[95] Ong (1958a, chapters 5 and 10).

[96] See Ong (1974), for Ramus's last emendations to his mathematics, made shortly before his death.

[97] Proclus (1992, p. 54): "Hippocrates wrote a book on elements, the first of whom we have any record who did so."

greatness of Pythagoras, and increased the stock of mathematical learning, improving them with an *Elements* that had a more complete and richer order and method.[98]

Ramus made a number of assumptions here. He imagined that Pythagoras and Hippocrates were both teachers, whose primary concern was to impart mathematics to their students – in accordance with his earlier portrayal of Pythagoras the schoolmaster. Hippocrates surpassed Pythagoras not in any specifically mathematical way, but in devising a better textbook. Ramus (in another unspoken, probably unconscious assumption) connected the compilation of mathematical *Elements* with the *teaching* of mathematics. It would be possible to attribute very different motives to Hippocrates: that he wanted to gather all known, fundamental theorems for the reference of practising mathematicians, for instance.[99] But this did not occur to Ramus. Hippocrates's book of *Elements*, like that of Euclid, was a textbook, and its success was to be judged by criteria like clarity and order.

Thus Ramus cast Hippocrates as a teacher much like himself, concerned with producing methodical textbooks for his students. Like Ramus, too, Hippocrates was unafraid to criticize his elders. Hippocrates (in his reading) was not scared off by Pythagoras's reputation; to the contrary, he boldly surpassed him, producing a better version of the *Elements*. There is, again, a clear analogy with Ramus, who criticized the ancients (including Euclid, the elementator *par excellence*), always to the end of surpassing them as a teacher. Hippocrates showed no arrogance in surpassing Pythagoras, only a kind of filial piety. In this and his subsequent remarks on later elementators, Ramus provided cover for himself against charges of *odium* and *invidium* for his own exuberant "correction" and "emendation" of Euclid (to say nothing of the suggestions of treason that Charpentier had raised).[100]

Ramus reimagined Hippocrates not only because he wished to suggest a resemblance between himself and the ancient elementator. His surpassing of Pythagoras was, in Ramus's historiography, a *normal* event: the story of Hippocrates contributed to Ramus's narrative of mathematical progress. Charpentier, on the other hand, rejected the very possibility of mathematical progress. In his third, victorious oration against Ramus, he had associated the desire to surpass the ancients with the envy and arrogance peculiar to Ramus and his followers.[101] In the *Scholae*

98 *Scholae mathematicae* p. 10: "Primus mathematicae in schola magister Pythagoras fuit, sed ut de primis initiis credi par est, minus distinctus, ut *stoikheiôtês* ideo non appelletur: sed tamen quidquid sit, Hippocrates Pythagorae magnitudine minime deterritus mathematicum magisterium auxit et exornavit elementis ordine, viaque pleniore et uberiore deductis."

99 Knorr connects Hippocrates's systematization of geometry with the problem of squaring plane figures. In his view, Hippocrates was concerned to catalog the techniques already known for squaring rectilinear figures, in order to narrow down the approaches to squaring curvilinear figures, especially lunules (of which Hippocrates squared three of the five quadrable types) and the circle itself. Knorr (1986, pp. 40–41).

100 Ramus wrote of Theudius, the third elementator in Proclus's catalog, that he "did not consider it odious or invidious to correct the *Elements* of Pythagoras and Hippocrates, or of Leon." (Ramus 1569, p. 19: "... Theudius ... nec odiosum sibi, nec invidiosum putavit Pythagorae, Hippocratis, Leontisque *stoikheiôsin* corrigere et emendare.")

101 Charpentier (1566, sig. G2r).

*mathematicae*, Ramus responded to Charpentier's criticisms historically, showing that the historical record bore witness to undeniable mathematical progress – Hippocrates had surpassed Pythagoras. Pythagoras's famous elation over his discovery of his eponymous theorem (offering a hecatomb of cattle in sacrifice) was intelligible only if new discovery was possible.[102]

The chains of elementators that Ramus listed repeatedly in the *Scholae mathematicae* were intended to reinforce this model of intellectual progress. Pythagoras, as a proto-Ramist mathematician, belonged in this main narrative even if, in the passage cited above, he could only be claimed as a writer of textbooks, not a fully fledged elementator – at least, not yet. For even as Ramus admitted Pythagoras was not an elementator, he attributed to him a substantial written mathematical work that looked very much like a collection of elements. Hippocrates was (according to Ramus) building upon and improving some sort of Pythagorean mathematical record, though he was deliberately vague about its precise nature ("whatever the case may have been").

This was the starting point of Ramus's assimilation of Pythagoras into the line of elementators. In his next significant reference to Pythagoras, he wrote that Leon (Proclus's second elementator, successor to Hippocrates) was the "*third* master and teacher of mathematical philosophy, and also a writer" who surpassed both Pythagoras and Hippocrates in his attention to utility.[103] Here, Ramus has placed Pythagoras first in a series of mathematical *writers*, if not elementators. He clarified that Leon was only the second *elementator*; but only two pages later he wrote that Theudius (the third elementator listed by Proclus) corrected the *Elements* of Pythagoras, Hippocrates and Leon,[104] now unambiguously attributing the first written *Elements* to Pythagoras. Shortly after that, he wrote:

> Pythagoras, if I may also count him as if he were an elementator ... Hippocrates emulated his fame, by writing down and publishing an *Elements* furnished with demonstrations.[105]

By the beginning of book III of the *Prooemium*, the transformation of Pythagoras was complete. Here Ramus included Pythagoras without comment at the head of the list of elementators, as if it were an established historical fact.[106] However, it must be repeated, there was no historical evidence that Pythagoras wrote a collection of *Elements*, nor any other mathematical text. Ramus was not simply making an error when he included Pythagoras in a lineage of mathematical authors. It took

---

[102] Ramus stressed that mathematical progress was continuous, by noting that Pythagoras himself was unaware of the more general, superior theorem that became *Elements* VI.31; if Pythagoras's theorem was worth the sacrifice of a bull, then, Ramus thought, the anonymous VI.31 deserved at least a thousand (Ramus 1569, p. 7).

[103] Ramus (1569, p. 17): "Leo igitur tertius mathematicae philosophiae non solum magister et doctor, sed scriptor Pythagora et Hippocrate usus laude perfectior et accuratior fuit."

[104] See n. 100 above.

[105] Ramus (1569, p. 19): "Pythagoras, ut hunc etiam tanquam *stoikheiôtên* numerem ... Hippocrates istam laudem aemulatus, elementa demonstrationibus exornata descripsit et publicavit."

[106] Ramus (1569, p. 77).

considerable effort and special pleading to establish him in this newly imagined role, and such a labor was not undertaken on a whim. If Pythagoras was to be the model Ramist schoolteacher, then he *must* have been concerned to present his material as clearly and "methodically" as possible – and therefore he must have written a textbook. It was only fitting that an ancient Ramist mathematician should also confirm Ramus's model of progress. His sympathy, even identification with Pythagoras is evident; but Ramus also praised those who inevitably surpassed Pythagoras. In this way, he not only highlighted the possibility of mathematical advancement, but also justified through history itself his own critical stance towards the mathematical past.

When he invented the Pythagorean "Elements" out of little more than thin air, Ramus pushed back the beginnings of recorded mathematics to a primitive era, in which the art was much closer to its original (and hence natural) form. In particular, Pythagoras wrote his *Elements* before Plato, whom Ramus now blamed primarily for the theoretical and demonstrative turn in mathematics and its subsequent decline, as we have seen. Mathematics *before* Plato was thus more authentic, although the seeds of its later corruption could be found even in the earliest period. Proclus recorded, for instance, that Hippocrates was the first to use reductions to the impossible. For Ramus (as for other Renaissance writers on mathematics) indirect proof was much inferior to direct demonstrative proof.[107] He interpreted Proclus's statement to mean that Pythagoras and other early mathematicians must have had direct proofs for their theorems, which indicated the nature of the thing itself, rather than merely persuading that it cannot but be the case, *per accidens*; and so Hippocrates, Pythagoras's successor in the chain of elementators, introduced a flaw into the *Elements* and (at least, in this respect) left the text in a poorer state than that in which he had received it.[108]

Thus Pythagoras's supposed *Elements* emerged during the natural, relatively unspoilt period of mathematical activity before the advent of Platonism, and even before one of the first significant departures from mathematical simplicity, demonstration by reduction. Despite its loss and effacement by later, decadent *Elements*, Pythagoras's textbook represented the possibility of a mathematics that taught directly, by showing rather than proving – the mathematics Ramus himself attempted to reconstruct in his own *Arithmetic* and *Geometry*. In yet another sense, then, Pythagoras was a model mathematician for Ramus; Ramus's own *Geometry*, though never completed to his own satisfaction,[109] can nonetheless be seen as an attempt to recover a pre-Hippocratean mathematics.

Finally, and quite surprisingly, Ramus found in Pythagoras a kind of "earthy" primitivism, a sort of mathematics in the body. This is paradoxical, to say the least, given Pythagoras's reputation even in the Renaissance for abstraction and mysticism. Throughout the *Prooemium mathematicum*, Ramus was repeatedly drawn back to the image of Pythagoras sacrificing a bull, or a hecatomb of bulls, in celebration of his discovery of his famous theorem. The first time he mentioned

---

[107] See Goulding (2005).

[108] *Scholae mathematicae*, p. 96.

[109] See Ong (1974).

this, he added:

> The loves of mathematics are at first bitter and difficult, yet eventually filled with pleasure.[110]

This lover's lament reflects Ramus's own passionate relationship with mathematics. We recall that his first prolonged encounter with the *Elements* brought on acute back-pain, halfway through book ten, at which he

> threw away [his] drawing-board and ruler, and burst out in rage against mathematics, because it tortures so cruelly those who love it and are eager for it.[111]

Pythagoras, and Thales before him, made grand, sacrificial gestures, carried away by their bitter love of mathematics. Eratosthenes would later put up a votive tablet for the same reasons; and Archimedes, by running naked through the streets, sacrificed his body and soul, his very reputation among his uncomprehending fellow citizens.[112] Perhaps Ramus's cramp in the spine does not compare with the sacrifices made by these legendary mathematicians. But, at the peroration of the *Prooemium mathematicum*, Ramus promised to repeat Pythagoras's sacrifice, a hecatomb of cattle to anyone who could

> make mathematics easy for boys, accessible to ordinary working men, and not only marvelous to know and use, but popular.[113]

Ramus's admiration, and physical sacrifice, in other words, was reserved for those who could make mathematics itself more earthy and physical. Pythagoras, who made such a grand physical gesture on discovering a theorem about a triangle, who examined his prospective students through their physical features, who wrote the first elementary textbook of geometry, showing the way directly and simply to the truths of mathematics – *this* Pythagoras, constructed in Ramus's historical imagination, was a fitting model for the reformed physical mathematics that Ramus himself sought. A great Ramist systematizer of mathematics would in fact be a second Pythagoras (since, as Ramus had shown, Pythagoras was in some sense a *first Ramus*). It would be entirely appropriate if he were honored with a Pythagorean sacrifice. Such a reformation of mathematics as a whole deserved more celebration than the discovery of a theorem in geometry, even one as fundamental as Pythagoras's. For by introducing Pythagoras's *method* in teaching and mathematical presentation, more than any actual geometrical results, the University of Paris would be refounded on Pythagorean principles.[114]

---

[110] Ramus (1569, p. 7): "Amores nempe mathematici sunt illi acerbi primum difficilesque, tandem voluptatis plenissimi."

[111] Text at n. 49 of second chapter above.

[112] *Scholae mathematicae*, p. 32.

[113] *Scholae mathematicae*, p. 112: "a quibus mathematicas artes pueris faciles, opificum vulgo familiares, cognitione denique et usu non tantum mirabiles, sed etiam populares factas esse videam."

[114] [scholmath] p. 13: "Ergo Pythagoras Academiae Parisiensi mathematicas optabit: Ergo Plato in Academia Parisiensi mathematicas artes desiderabit; et uterque Parisiensem Academiam, tum Pythagoream et Platonicam esse judicabit, cum mathematicis primas in philosophia detulerit."

One can admire the thoroughness of Ramus's discussion of Pythagoras, if not always his historical reliability. The result of his meditation on the ancient figure was a vision of mathematics as a living discipline – one that could be realized again, if only the Ramist curriculum of studies were implemented in the modern university. This is a reformer's manifesto written in the language of history, much more compelling than the rhetorical and polemical modes he had previously deployed. Considering Ramus's increasing marginalization in the university after his defeat over the Charpentier chair, there is something almost poignant about his discovery of an authentic mathematical school, flourishing under his own pedagogical principles, deep in the legendary past.

## Ramus and Charpentier, and the Mathematization of Nature

When Ramus claimed Pythagoras as his own in the *Scholae mathematicae*, he also assigned Charpentier a new historical role: that of Aristippus. After his appearance in Melanchthon's prefaces, Aristippus had become a well-known figure in Renaissance introductions to mathematics. In Melanchthon's version of the shipwreck anecdote, the pleasure-addicted hater of mathematics was humbled and forced to recognize mathematics as a civilized, liberal pursuit.[115] In his defense of mathematics against the charge of *inutilitas* in the second book of the *Prooemium mathematicum*, Ramus referred all criticisms that mathematics was useless to "Aristippus," a transparent sobriquet for Charpentier. Aristippus, wrote Ramus, teaches Aristotelian physics,[116] but is so unaware of its mathematical basis that he is like a blinded Polyphemus, his classroom a kind of Cyclops's cave.[117] This Aristippus, like his historical namesake, criticizes mathematics for saying nothing about the good and the beautiful; for such obtuseness he would be thrown out of any decent university, such as any of those in Germany.[118] He is a *varius homo*, a chameleon, who one moment is a scholastic, and the next plays the courtier. He will slander mathematics, much to the displeasure of his patron;[119] and then, when he must, he will adopt a feigned philosophy and sing its praises.[120]

These last observations led Ramus to the story of Aristippus in the shipwreck.[121] In Melanchthon's version, Aristippus had always secretly acknowledged that mathematics was a civilized art, even though he admitted it only *in extremis*. Charpentier,

[115] See passage quoted at n. 61 of first chapter.

[116] *Scholae mathematicae* p. 46.

[117] *Ibid.*, p. 49.

[118] *Ibid.*, p. 71.

[119] That is, the Cardinal of Lorraine, who was Ramus's patron until Ramus converted to the reformed religion. Throughout the case, even as Charpentier mocked him for having lost his patron, Ramus affected that he and the Cardinal remained close.

[120] *Ibid.*, p. 74.

[121] *Ibid.*, p. 75.

the hater of mathematics angling for a chair in the subject, could clearly be represented by this story; but such a charitable reading of Aristippus's motives would hardly suit Ramus's purposes. Instead, continuing his earlier characterization of Charpentier as fickle, Ramus quoted a verse from Horace, that "Aristippus could accommodate himself to every condition, rank and circumstance."[122] In that poem, Horace related how Aristippus was teased by a Cynic for his enjoyment of royal luxury. The Cyrenaic replied that he worked hard entertaining the king, and received regular reward; his critic had to beg and receive less than he did. Ramus's Aristippus, then, was a paid lapdog of the gentry, presenting himself for the chair of mathematics only to win his masters' admiration and patronage. The use of the legend allowed Ramus to make the charge obliquely without offending Charpentier's powerful backers themselves. (Perhaps, too, the rather pathetic figure of the Cynic in Horace's story was Ramus himself, aware that his defeat and loss of patronage had reduced him to begging scraps at the tables of the great).

Charpentier, as one might expect, bristled at his characterization as a latter-day fawning Aristippus, and in his *Admonitio ad Thessalum*, written in reaction to the *Prooemium*, he rejected the title angrily. The name-calling had escaped the bounds of Ramus's book; Charpentier reported that Ramus's students had begun to use the name in public, accusing Charpentier of following Aristippus in his belief that the mathematical arts have no goal.[123] This, he claimed, was not true, although if it were true he would simply be repeating the opinion of Plato and Aristotle: that mathematics has no *practical* goal. Charpentier then turned to the second book of the *Prooemium mathematicum*, where Ramus had so vigorously attacked him in the figure of Aristippus. But despite Ramus's claim that mathematics was, indeed, useful, Charpentier could discover in Ramus's work no "end" of mathematics worthy of the name: only base, illiberal applications. Charpentier insisted that the only application of mathematics was that which Proclus, Plato and Aristotle had proclaimed (whether Ramus liked it or not): the elevation of the soul to mathematical objects, which lay at the midpoint between sense objects and the entirely immaterial.[124]

Ramus, of course, was never going to accept this correction, least of all from Charpentier; he had spent decades writing and teaching about a mathematics that was grounded in the real, physical world. Mathematics was to be the foundation of the other sciences, not because it trained the mind to think more precisely (a position that few would have criticized), but because it described the natural order of the world and the mind – the same order that the other arts reflected. And so, in the most fascinating twist that the debate would take, Charpentier turned his attention to the very idea that mathematics could underlie the other arts, scrutinizing a passage in the *Prooemium mathematicum* that hit close to Charpentier's own field of natural philosophy.

[122] Horace, *Epistulae* I.17: *Omnis Aristippum decuit color et status et res.*
[123] Charpentier (1567, fol. 18v).
[124] *Ibid.*, fol. 22r.

Ramus argued (or asserted) that "Aristippus" was unqualified to teach the natural philosophy of Plato or Aristotle, because every aspect of the philosophers' physics was grounded in mathematics.[125] The notion of a mathematical physics is, of course, very significant for the emergence of early-modern science. Ramus was led to this position in part by Melanchthon's arguments; indeed, many of the examples of mathematical physics he provided were the same as Melanchthon's.[126] but also, in part, simply out of a desire to discomfit Charpentier and other teachers of Aristotelian natural philosophy.

Ramus had no difficult making the case for Plato: the mathematical foundation of the *Timaeus* spoke for itself. The case for Aristotle, on the other hand, was rather more *ad hoc*. Ramus recalled that the philosopher used frequent examples from the geometers throughout his writings, relied on geometry to explain the rainbow in the *Meteora*, and in his *Physics* dealt with rest and motion, a subject he treated more precisely and mathematically in his *Mechanics* (a work that Ramus himself taught to his students). This would hardly be sufficient to maintain that Aristotle's physics was *essentially* mathematical, but for Ramus it sufficed. And the mathematization of physics could, he inevitably claimed, be traced back to the Pythagoreans: the axiom of physics that all things move to their place of rest at right angles, for instance, was intimated by the Pythagoreans in their making the figure of earth a cube.

Charpentier made short work of Ramus's Pythagorean, mathematical physics. In doing so, he revealed some fascinating indications of the direction Ramus's thought was taking after the publication of the *Prooemium mathematicum*. His attack also exposed just how difficult it was for anyone at this time (including Ramus) to imagine what a mathematized world might actually look like. Ramus had insisted that the Pythagorean school had based their physics on mathematics, and that Aristotle had followed their example. In reply, Charpentier insisted on a distinction between historical and philosophical truth. He allowed that Ramus might well be right to claim that Pythagoras (if not Aristotle) explained the world in a mathematical way. There seemed to be no doubt, *historically*, that the Pythagoreans explicated nature by numbers and figures. But Ramus had misconstrued the *philosophical* significance of their position, and thereby misinterpreted the significance of their actions. The ancient philosophers only described nature in the language of mathematics because numbers and geometrical figures had an almost proverbial *obscurity*. As Charpentier went on to explain, Pythagoras and his followers used mathematics as a kind of veil or cipher to keep the vulgar away from their philosophy; only those initiated into Pythagoras's own geometrical teaching were able crack the code and access the truths hidden beneath the geometry and arithmetic.[127] Charpentier professed himself frankly bewildered that Ramus should find such an approach laudable. And, even if Plato had been guilty at times of resorting to the same obscurity in the *Timaeus*, it

[125] Ramus (1569, pp. 46–47).

[126] See p. 14 above.

[127] Charpentier (1567, fol. 58r): "Sicque Pythagoreorum institutio et paedia, erat posita in mathematicis, quoniam, ut dixi, Philosophiae mysteria, quae volebant a suis tantum intelligi, illi per numeros et figuras explicabant."

was ludicrous to claim that Aristotle had similarly engaged in deliberate obfuscation.[128] If, as some have argued, the new philosophers of the seventeenth century saw nature as a code waiting to be cracked,[129] an Aristotelian like Charpentier saw the mathematical explanation of nature as the *imposition* of a cipher onto a plaintext world.

It may be that Ramus had been led to a mathematical view of nature in part to set himself apart from his Aristotelian opponent. And he had, of course, always seen mathematics as primarily directed to practical use in the world; the mathematical natural philosophy of the *Prooemium mathematicum* was, in a sense, a pointed restatement of that fundamental conviction. But his long meditation on the figure of Pythagoras and his teachings seems to have turned him towards a quite original position, namely that mathematics was not only useful, but it was also the language, the very *substance* of the physical world. Charpentier reported that Ramus's lectures on geometry, delivered after the *Prooemium mathematicum* (and never collected or published) had taken a bold new turn:

> Or is it that you would take refuge in something that (as I hear) you recently maintained in a public lecture on geometry? You said that the subject of arithmetic and geometry (quantity, in other words) is not some accident of that substance which constitutes a natural body, but is in fact the natural body's principle and foundation itself. Now, even though many people I trust told me that you said this several times in your lecture, and even though I can guess where you want to go with this opinion, at least as far as religious questions are concerned, still, this seems to me so absurd and so monstrous an opinion that I simply wouldn't dare to ascribe it to anyone, even to you.[130]

If Charpentier's report is to be trusted, Ramus eventually achieved complete identification with Pythagoras. Only a year before, he had been entirely indifferent towards the ancient philosopher. Then, he remade the Pythagoras in his own image, carefully avoiding any hint of Pythagoras's obsession with numbers. Finally, (according to Charpentier) Ramus subsumed even that previously unpalatable feature into his worldview.[131]

The mathematical foundations of physics would become a central issue in the development of natural philosophy, within just a few years of the Ramus-Charpentier

---

128 *Ibid.*, fols 58r–59v.

129 See Pesic (1997).

130 Charpentier (1567, 60r–v): "An ad id confugies quod audio nuper in explicatione geometriae tibi factum esse familiare? Arithmeticae scilicet et geometriae subiectum, quod quantum dicitur, non esse affectionem eius substantiae ad quam naturale corpus refertur, sed eius principium atque fundamentum. Equidem etsi permulti fide dignissimi, mihi testati sunt, hoc a te saepe in tuis praelectionibus esse praedicatum facileque suspicer hac nova opinione quorsum in his quae ad religionem pertinent velis evadere, haec tamen mihi tam absurda est tamque monstrosa, ut non audeam tibi eam hoc loco ascribere." The religious implication was the denial of substance and accident, which would undermine the doctrine of transsubstantiation of the elements of the Eucharist.

131 Traces of Ramus's late lectures on Pythagoras may be found in his *Scholae metaphysicae* and *Scholae physicae*, where his analysis and critique of the third and sixth books of the *Physics* and the tenth book of the *Metaphysics* (the latter Aristotelian text one of the classic attacks on Pythagoreanism) led him to a kind of Pythagorean atomism.

debate. The two non-mathematicians involved in this conflict explored the subject of the mathematical basis of the sciences largely through the historical imagination. In order to defend mathematics, Ramus tried to imagine a time and a situation in which mathematics was not just useful, but central to the teaching of all the arts. He had been concerned with mathematics and its connection to the world since his very first writings; but, in this debate, his fixation on the person of Pythagoras as a founder both of mathematics and of the Ramist method in the arts, drew him closer to the ancient philosopher's mathematical realism. Charpentier's opposition to Ramus over his mathematical physics was, in one sense, philosophical and religious, but his central point was also historical: that Pythagoras used mathematics to conceal knowledge, not to convey or discover it. There is some irony in the fact that Charpentier, although often a better historian than Ramus, was here defeated by a lack of imagination. Ramus's construction of Pythagoras, on the other hand, was a piece of thoroughly partisan historiography. But, by his reimagining of the mathematical past to suit his polemical ends, Ramus inadvertently stumbled on a most fruitful path for the future development of the sciences.

## The Royal Road

Pythagoras became a vehicle for Ramus to work through concerns about mathematical pedagogy. In another case – an incident in the life of Euclid – Ramus again used an ancient mathematician to stake out the correct form of mathematical instruction, portraying him, in fact, as if he were in a debate over this very subject. One of the very few pieces of (apparently) substantial information about the life of Euclid was preserved in Proclus's *Commentary*: asked by Ptolemy I, king of Egypt, whether there was any quicker way to master geometry than the *Elements*, Euclid replied that there was "no royal road to geometry."[132]

This anecdote would be used by many Renaissance historians, including Ramus himself, to locate Euclid in time and space.[133] For Ramus, however, the *substance* of the remark also possessed great significance. In fact, the "royal road" was freighted with great meaning in the *Prooemium mathematicum*, a shibboleth for Ramus's dissatisfaction with ancient mathematical pedagogy. On the very first page of the *Prooemium*, as he set out the plan of the work, Ramus wrote that the third book would "argue the problem of King Ptolemy against Euclid, concerning the easier and quicker way of teaching mathematics."[134] As Ramus interpreted it, Proclus's anecdote was much more than the record of a scholar's clever rejoinder to a king; it indicated a dispute between Euclid and Ptolemy over the nature of mathematics itself.

---

[132] Proclus (1992, p. 57).

[133] See fifth chapter.

[134] Ramus (1569, p. 1): "Tertius, Ptolemaei regis problema adversus Euclidem disputabit, de magis perspicua, magisque compendiaria via matheseos instituendae."

Ramus reprised this anecdote when recounting the life of Euclid later in the *Prooemium*. He expanded the encounter between king and geometer far beyond Proclus's brief notice. Ptolemy, he tells the reader, had heard report of Euclid's geometry,

> but that king does not seem to have approved of the reasoning or the route taken by Euclid's *Elements*. And Euclid himself does not seem to have acted very generously towards the king. For, it is said, the king once asked Euclid whether there was a shorter route to geometry than that of the *Elements* he had composed. Euclid replied to him, "Sire, there is no royal road to geometry." By saying this he meant, it seems, that the road of the *Elements* that he had composed was broad, open, simple and straight – a military road, if you like, and thus fit for a king. But he also meant that any shorter path would be slippery and treacherous, and therefore not fit for a king. But in the third book the problem will be considered more fully: whether the judgement of the king or of Euclid was more reasonable in this matter.[135]

Euclid's disavowal of a "royal road" seems to be just that: a denial that there was any *special* path to geometry for Ptolemy, and an insistence that all seekers after geometrical knowledge – from schoolchildren to kings – must start at the same level and proceed by the same, at times painful steps towards mastery. But this is not how Ramus interpreted this anecdote. According to Ramus, Euclid meant that his own way (the *Elements*) was itself the royal road, the only path suitable for a king, and that any *other* path would not be kingly. Ramus's repeated emphasis here on the *Elements* "that [Euclid] had composed" left no doubt that it was Euclid's particular *version* of geometry that was in question, not the art itself. Ramus found Euclid's reply to the king neither clever nor witty, but "ungenerous," exhibiting the same kind of self-pride and preference for one's own creation over the natural art that he blamed for the decline of all arts and sciences.

The encounter between Euclid and Ptolemy thus became more than an example of a scholar's wit. There was a real question at stake in the conversation between king and geometer: what *is* the royal road to geometry? Ramus's Euclid pronounced that his *Elements were* the answer to the king's question. In doing so, he made a substantive claim about mathematics: that the only "broad, open, simple and straight" route to geometry was the *Elements* as he had recorded them. Ramus did not find Euclid's self-assurance "reasonable" in any way. The implication that Ramus wanted to draw was that there was indeed a royal road to geometry, but Euclid's work had strayed far off that path.

In another passage, Ramus compared Euclid to Plato in his baleful guise as ruiner of mathematics. Euclid's gloomy refusal to countenance any easier and more direct route to geometry, the overweeningly proud manner in which he turned down the

---

[135] Ramus (1569, p. 24): "*stoikheiôseôs* tamen Euclideae rationem et viam videtur rex ille non probasse, neque Euclides ipse satis liberaliter regi fecisse. Rex enim Euclidem aliquando interrogasse fertur, num qua ad Geometriam via magis compendiaria esset, quam *stoikheiôseôs* ab eo compositae; cui Euclides, Semita (inquit) ô rex, ad geometriam regia nulla est; quo responso videtur significasse viam elementorum a se compositorum esse latam, apertam, simplicem, directam et tanquam militarem, ideoque regiam esse. Semitam autem breviorem esse lubricam et ancipitem, neque ideo regiam. Sed istud problema tertio libro plenius edisseretur, Regisne hac in re judicium, an Euclidis *logikôteron* fuerit.

request of a king – it all paralleled Plato's proscription of mechanical methods in geometry. Both men refused to release knowledge into the hands of ordinary laymen, whether kings or common artisans.[136] As Ramus had argued, this growing distance from everyday needs and concerns was the source of mathematics' fabled "obscurity;" Ptolemy's inquiry was intended as a rebuke to Euclid and a reminder of the dangers of abandoning the clarity of application.[137]

Ramus returned to this anecdote repeatedly throughout the *Prooemium mathematicum* (and even in the additional books of the *Scholae mathematicae*). For instance, at the end of his devastating (he thought) critique of Euclid's logic in the third book of the *Prooemium mathematicum*, after he had shown to his own satisfaction that the *Elements* was riddled with faults and all but unusable as a school text, Ramus returned to the confrontation between Euclid and the Egyptian king:

> And now, let us settle the whole case arising from the king's complaint. Ptolemy objects that Euclid's *Elements* are obscure and difficult. Euclid, on the other hand, maintains that they are clear and easy – so much so, that if anyone wanted a clearer and shorter one, he would be seeking a slippery path, not a royal road. Proclus appears for the defense, on behalf of Euclid's *Elements*. I have taken up the king's cause, and to this point have conducted it according to the laws agreed upon by the consent of all parties.[138]

Again, Ramus interpreted the anecdote as a positive claim by Euclid that *his* was the royal road. But here he employed the metaphor of a litigation, in which Ramus himself would take the king's part. These words are quite striking, given the circumstances in which he wrote the *Prooemium*. This polemical history, addressed to the Queen Mother Catherine de Médicis, was published in the aftermath of a devastating legal defeat. In the courtroom, Ramus had been practically accused of treason and disloyalty to the crown, while his opponent Charpentier basked in the patronage that Ramus himself had once enjoyed, and secured for himself the award of the regius chair. Yet, Ramus assured his reader and his royal audience, he himself would be the advocate for the crown's case, because there *was* a royal road to geometry: Ramus's direct and practical mathematics.

[136] *Ibid.*, p. 28.

[137] *Ibid.*, p. 79.

[138] *Ibid.*, p. 104: "Quapropter totam regii problematis querimoniam concludamus. Ptolemaeus queritur Euclidis *stoikheiôsin* obscuram et difficilem esse; Euclides contra confirmat esse perspicuam et facilem, ut si quis illustriorem aut expeditiorem requirat, lubricam semitam quaerat, non viam regiam. *Stoikheiôsis* Euclidis defenditur a Proclo, caussa regis a nobis suscepta et hactenus secundum leges consensu partium laudatas ac probatas acta est." Proclus's defense is, of course, the *Commentary*, which Ramus claimed supported in principle the three laws of Ramist method – laws that he argued Euclid had violated (hence the reference to "laws agreed upon by the consent of all parties.")

# Chapter 4
# "To Bring Alexandria to Oxford:" Henry Savile's 1570 Lectures on Ptolemy

## Introduction

The most substantial reaction to Ramus's histories of mathematics came from an unexpected quarter. In September 1570, a young Oxford master, Henry Savile, began to deliver a series of ordinary lectures on astronomy – "ordinary" because they were structured around the orderly reading of a text, but in every other respect quite out of the ordinary. They brought the 20-year-old lecturer local fame, and were to be remembered as one of the notable academic events in Elizabethan Oxford. Nearly a century later, the antiquarian Anthony à Wood described how

> our author Savile proceeded in his faculty, and read his ordinaries on *The Almagest of Ptolemy*: whereby growing famous for his learning, especially for the Greek tongue and mathematics (in which he voluntarily read a lecture for some time to the academicians) he was elected proctor of the university for 2 years.[1]

This was only the beginning of an illustrious academic career. Savile would eventually serve as both Warden of Merton College and Provost of Eton. At the end of his life, he transformed the study of mathematics in England by establishing the Savilian Chairs of Geometry and Astronomy at Oxford, the first professorships in any mathematical subject in England. With this benefaction, made from a position of great authority, Savile finally fulfilled the lofty purpose he had set out in his ordinary lectures 50 years before: to "restore mathematics to the university."[2] But that was all still to come; in 1570, as a young, unknown lecturer, Savile confronted a university almost completely indifferent to the study of mathematics. In his ordinary lectures, he did more than just expound on the content of Ptolemy's text. He also set out to demonstrate to his students the philosophical nobility of the mathematical arts. To do this, he would employ every device in the canon of classical epideictic rhetoric, from a stirring protreptic praising the honesty and utility of the quadrivium to a long series of historical exempla illustrating the heroic feats of the mathematical

[1] Wood (1813–1820, p. 310).

[2] MS Oxford, Bodleian Library, Savile 29, fol. 2v: "Sed O me somnio nescio quo felicem, qui hoc seculo, his hominum moribus scholis mathemata mathematicis dignitatem restituere sperarem."

R. Goulding, *Defending Hypatia*, Archimedes 25,
DOI 10.1007/978-90-481-3542-4_4, © Springer Science+Business Media B.V. 2010

practitioners of antiquity. By retracing the history of mathematics, Savile sought to both legitimate and define the discipline. His efforts followed closely on the work of Ramus, whose *Prooemium mathematicum* served Savile as a major source of information and intellectual inspiration, as I shall show. In the end, however, Savile would direct Ramus's project of mathematical history to a dramatically different conclusion.

## The Making of a Mathematician

Savile was born in 1549, of a middle-class Yorkshire family.[3] At the age of 12, he matriculated at Brasenose College, Oxford (where his father had been a student) and 4 years later, in 1565, before even obtaining his bachelor's degree, he was elected to a Fellowship at Merton College, the college with which he was to be associated for the rest of his life. He graduated BA the following year, and embarked upon studies for his MA, which he received in 1570. His mathematical education can be reconstructed from the notes he made in the late 1560s, towards the end of his bachelor's degree and over the course of his MA studies, and from the books he read and annotated during the same period.

To all appearances, Savile was an autodidact, who had single-handedly mastered the mathematics of Euclid, Ptolemy, Archimedes and Copernicus.[4] His devotion to the *Almagest* extended to preparing a new translation of the text, with selections from ancient and Byzantine commentaries.[5] His guides through this difficult material were, above all, the textbook writers of the German universities – mathematicians like Erasmus Reinhold and Caspar Peucer.[6] Their writings not only provided Savile with an introduction to astronomy; they also offered a model for

[3] For Savile's biography, see Goulding (2004a); Wood (1813–1820, vol 2, pp. 310–317); Feingold (1984, pp. 124–131); Highfield (1997).

[4] Savile's heavily annotated copy of Euclid is in the Bodleian Library with shelfmark Savile W.9(1) (Euclid, *Stoicheia*, ed. Simon Grynaeus, Basel, 1533; this, the *editio princeps* of Euclid's Greek text, also contains the first edition of Proclus's *Commentary* on the *Elements*; Savile has also annotated this). The style of the hand shows that the annotations are certainly from Savile's early career. His copy of Archimedes is so copiously annotated that it was once classified as a manuscript (MS Savile 51); this Greek printed book (*Opera quae quidem extant omnia*, Basel, 1543) is now Savile X.9(1); it may have been annotated in part later in his career. I have not found Savile's copy of Copernicus.

[5] The translation occupies MSS Savile 26–28, and is dated to the first term of the 1568–1569 university year.

[6] One of the most copiously annotated volumes in the whole of Savile's library is his copy of Reinhold's translation and commentary of the first book of the *Almagest* (Savile Aa.13(1): *Ptolemaei Mathematicae constructionis liber primus, additae explanationes aliquot locorum ab Erasmo Rheinhold*, Paris, 1556. Savile clearly annotated the book while writing his own translation. Peucer's *Elementa doctrinae de circulis coelestibus, et primo motu* (Savile Aa.14; annotated by his younger brother Thomas, but not by Savile himself) served not only as a good elementary introduction to astronomy, but more importantly as a source for Savile's history of mathematics,

how the sciences ought to be taught in the English academies. The Germans taught astronomy through the committed study of a few central, highly-technical texts. This was a long way from the teaching of mathematics at Oxford where, Savile would later complain, teachers either relied on simplified handbooks or taught their students specific *applications* of astronomy, particularly the construction of horoscopes.[7] Savile's appointment as ordinary lecturer in astronomy in 1570 would allow him to put the educational ideas of his favorite continental practitioners to the test.

Through intense, largely independent study over the previous three years, Savile had taught himself mathematics and astronomy to an extraordinarily high level. At some point in his mathematical education, a copy of Ramus's 1567 *Prooemium mathematicum* had fallen into his hands – a work that was to prove both inspiring and infuriating to the young Savile.[8] In many ways, Savile's entire mathematical career can be understood not only as a campaign to restore the mathematical sciences at Oxford, but also as a continuous debate with the French logician. In his 1570 lectures, his first step in the promotion of mathematics, Savile not only brilliantly expounded the mathematics of both Ptolemy's *Almagest* and Copernicus's *De revolutionibus*, texts largely untaught in the English universities. He also delivered a stirring rhetorical defence of mathematics which emphasized the discipline's philosophical purity and ancient pedigree, while also attacking Ramus's promotion of mathematics as a practical science head-on.

The serious study of mathematics as part of the arts degree was no more popular at Oxford than it was at Paris. At Paris, resistance to Ramus's reforms came from Aristotelian philosophers who saw mathematics as at best useless, and at worst an attempt to interfere in the teaching of their field. (Such disciplinary trespassing was, of course, precisely Ramus's intention.) At Oxford, by contrast, literary humanism was in the ascendant, and its practitioners were simply indifferent to mathematics. The subject was certainly being studied privately and within the colleges of Oxford. It was in Oxford, after all, that Savile developed his interest in mathematics in the first place. The university was home to figures such as the mathematician Thomas Allen of Gloucester Hall, who tutored generations of Oxford pupils in arithmetic and geometry. But the art of mathematics had no public champion, no one to argue for its place in the university arts degree, as Ramus had done in Paris.

The champion who eventually did come forward was not an established professor, but a young graduate at the start of his career. In this, Savile was very different from Ramus, whose sustained defense of mathematics was initiated as a professor of the Collège Royale and with the support of powerful patrons. Moreover, in his *Prooemium* Ramus collected the lectures he had made as royal professor, issuing them as an appeal to the crown against Charpentier; his history was produced in

[7] See discussion below of Savile's lectures.

[8] Many of Savile's books and manuscripts eventually made their way into the library he established for the two mathematical professorships he founded at Oxford – a library that was eventually absorbed into Bodleian Library. Unfortunately it seems that he did not bequeath his copy of Ramus's *Prooemium mathematicum* to the library.

very public circumstances and to a specific political end. It is not surprising that Ramus produced a *tour de force*, sufficient (he hoped) for the gravity of the situation and the interest with which it was anticipated. But Savile's 1570 lectures were *not* a special series; they were merely ordinary lectures from which very little would have been expected, as I shall show. For this reason, it is all the more impressive that Savile's work is at least the equal of Ramus's, and often excels it.

Ramus's *Prooemium mathematicum* served as Savile's principal inspiration and most important (though often unacknowledged) source; and, in the long run, Savile was to shape mathematics at Oxford as much as Ramus did at Paris. But, even as Savile took Ramus as his model, borrowed information and material from his *Prooemium*, and shared his vision of a reformed university curriculum in which mathematics would be an integral part, nonetheless he took a fundamentally different view of the value and purpose of the mathematical sciences. Ramus sought to transform the humanities so that they would be more like mathematics; Savile argued that mathematics was in fact a humanistic art. It was to this end that he bent *his* history of mathematics, a lengthy, learned survey of the origins of the discipline from the sons of Adam to Ptolemy which dominates the early portion of the 1570 lectures.

Savile's debt to Ramus is clear. At the start of his lectures he wrote the title "Prooemium mathematicum,"[9] and throughout his text he lifted phrases, sentences and even whole paragraphs from Ramus's history, silently incorporating them into his own.[10] Even the rapture he expressed at the thought of bringing mathematics back to Oxford was copied almost word for word from Ramus's text.[11] Imitation may be the sincerest form of flattery; in this case, however, Savile plundered Ramus's words and thoughts only to turn them against him. In direct opposition to Ramus's practical mathematics, Savile vigorously maintained the more traditional view of mathematics as an abstract science, closely connected to Platonic philosophy. In support of his philosophy of mathematics, Savile reframed the historical narrative of Ramus's *Prooemium mathematicum* from the point of view of a theoretical mathematician – no easy task, since Ramus had originally crafted his narrative to underpin his *anti*-theoretical understanding of mathematics.

Like Ramus, Savile sought to defend mathematics against charges of obscurity and difficulty. Ramus had answered these attacks on mathematics by, on the

---

[9] MS Savile 29, fol. 2r. The title most probably covers the introductory section of the lectures, the description of the individual sciences and the history of mathematics (Sects. 1–3 of the division given below), thereby taking in all the material corresponding to Ramus's own *Prooemium mathematicum*.

[10] To give but one example, MS Savile 29, fol. 6v, starting at the phrase "Physica illa quae dicitur...," is closely based on *Scholae mathematicae*, p. 46 (see discussion of this passage at p. 69 above). It is ironic that Savile, asserting here that physics was essentially mathematical, apologized for the boldness of his claims; everything he said about the physics had in fact been said before by Ramus. A similar, but far more significant, example of Savile's pretended boldness while copying from Ramus (the redating of Euclid) will be examined in the next chapter.

[11] Compare text at n. 2 above with Ramus (1569, p. 110): "Sed o me somnio nescio quo tota cohortatione felicem et fortunatam! Sum P. Ramus regius Lutetiae professor ..."

one hand, emphasizing its practical utility and, on the other, castigating ancient mathematicians who had shrouded the primitive simplicity of the art in wreaths of demonstration and redirected it towards vague, contemplative ends. Savile, drawing on many of the same sources as Ramus, came to entirely different conclusions. Mathematics was not obscure, since it was directed to the perfect clarity of Platonic contemplation. Its difficulty arose from many factors, chief among them the mediocrity of modern university teachers and their reliance on simplified handbooks and summaries. Students emerged from their astronomy classes able to draw circles representing the planets, and little more; they had no idea of the reasoning that supported such celestial models, nor any understanding of the mathematics that they were built on. The remedy, Savile suggested, was to return to the texts of the ancients and to adopt the critical methods of contemporary philological humanism.[12] This was the approach that Savile himself took in his lectures; he would also enshrine this ideal in the statutes from his two professorships of mathematics, ensuring that for several generations Oxford mathematicians would combine mathematical research with antiquarian studies.

## Mathematics at Oxford

Was Savile right to lament the state of mathematics at Oxford, or was this merely a rhetorical claim? A review of university mathematics teaching at Oxford in Savile's time will shed some light on this question. In what follows, it should be noted that I am discussing only the teaching provided by the University of Oxford, not that which took place more informally in the individual colleges, where the state of mathematics may have been altogether more healthy. Nevertheless, the situation of mathematics in a uniformly taught, university-wide arts curriculum was a subject of concern for Savile and others.

In the sixteenth century, all university-level teaching in the arts was provided by newly created Masters of Arts. It had been a condition of the MA degree, from the very foundation of the universities, that the new masters should remain at the university for 2 years after inception, or award of the degree, to deliver "ordinary lectures." In these lectures, which were compulsory for undergraduate and MA students, the "regent masters" would read out a text set by the university and comment upon it.

The system, at least at the time it was founded, had some advantages for the university. First, the teaching cost the university nothing; the lecturers were paid directly by the students, each surrendering a shilling or two to his teacher at the start of the year's course. There was also a guaranteed pool of teachers at the start of each academic year.[13] The university thus kept itself at arm's length from the

[12] For a detailed account of Savile's diagnosis of the shortcomings of Oxford's system of mathematical instruction, see Goulding (2002) and Goulding (1999). The passage in which Savile considers the problems besetting mathematics at Oxford is considered below, at p. 83.

[13] Fletcher (1986).

provision of teaching; all of the expense and labor was borne by the undergraduates and graduates. But the strengths of this arrangement also created problems; by the late sixteenth century, the system had reached a crisis point.

In the early university, regent masters had been free to choose the art on which they wished to lecture. Inevitably, this led to an uneven distribution of teachers among the arts. To remedy this, a statute of 1431 stipulated that each year's new supply of regent masters should be divided into ten subject groupings (the seven liberal arts and the three philosophies, moral, natural and metaphysical); each regent master was permitted to lecture only on his allotted discipline.[14] Around the same time, the university and its benefactors built the Divinity School and Arts Schools in order to provide a central location for the delivery of ordinary lectures. Previously, the university had not even provided rooms for masters and students to meet; it was the teacher's responsibility to find a space – often only a shopkeeper's spare room – in which to deliver his lectures.[15]

But there were more fundamental problems in the system which were not so easily fixed. For some regent masters who aspired to a career outside of the university, 2 years spent in compulsory service was an unwelcome deferral of their ambitions. And for some, it meant virtual penury. While good teachers could attract enough students to make a decent living, those who were not so gifted found themselves lecturing to an empty room – satisfying the letter of the statutes but receiving no payment whatever in return.[16] It is not surprising, then, that the sixteenth-century university register records dozens of masters asking to be excused from their teaching duties, pleading the pressure of "business concerns" or illness.[17] Some simply refused to satisfy their obligations; on these, the university imposed fines, which must have seemed a small penalty to avoid the 2-year burden of teaching.[18]

The situation was exacerbated by the unwillingness of many students to attend the ordinary lectures. They were *obliged* to attend, and a fine was levied on those who skipped them; but the long lists in the register of fines received from undergraduates show that the students were as reluctant to learn as the masters were to teach. The fault, it seems, lay partially with the form of the teaching itself. Before the advent of printing, it had made perfect sense for a master to read out a manuscript of a set text and append his own observations upon it. But when printed texts were readily available, along with commentaries by noted scholars, this method of teaching must have seemed tedious and irrelevant.[19]

---

[14] Gibson (1931, p. 235).

[15] Harvey (1992, pp. 750–751). A document from 1300 records 54 such "schools" scattered throughout the city of Oxford; see Pantin (1972, p. 235).

[16] On these so-called "wall lectures," see Mallet (1924, vol. 1, p. 199).

[17] Fletcher (1986, p. 186); Clark (1887–1889, vol. 1, pp. 96–99).

[18] A decree condemning masters and students who missed lectures was issued in 1556–1557 (Gibson 1931, p. 369). Another decree, of 1566–1567, laid down a fine of a shilling for each lecture omitted by an ordinary lecturer (*ibid.*, p. 398).

[19] Fletcher (1986, p. 187).

There were also deeper problems besetting the ordinary lectures, reflecting large intellectual and even demographic changes at the two English universities. Mark Curtis has argued that the universities of England went through a revolution in the sixteenth century in their institutional structure and curriculum. The reason was simple: the gentry had discovered the university. Before the mid-sixteenth century, the only aristocracy who attended the university were those who intended a career in the church – a relatively small number. With the spread of Renaissance ideals from the Continent, education in the humanities became more and more essential for the ideal gentleman. At first, such studies were pursued not at the academy, but under the guidance of a private tutor. Early English humanist writers on education viewed the universities as providing the very *opposite* of a liberal education.

By the late sixteenth century, however, there were so many wealthy gentlemen at the universities that clergy complained that poor scholars destined for the church were being squeezed out; the universities themselves were aware of the stratification of their student body, and vainly attempted to address the problem by limiting extravagance in dress. The universities had become more attractive to the gentry for several reasons. From the early sixteenth century, masters and professors at the university, themselves influenced by the new learning from the Continent, had begun to integrate humanist texts into the university curriculum – a process that was accelerated by the influx of students seeking a humanist education. The less wealthy of the gentry discovered that a humanist education could be had at the university that was much *cheaper* than a private tutor. As students spent more and more time within their colleges, under the academic and moral charge of a college tutor, the universities came to seem the most attractive option for the education of well-born boys, all but indistinguishable from the sort of education formerly provided at home, by private tutors.[20]

Both universities reacted to these dramatic changes by altering the syllabus to accommodate humanist texts. Even the structure of the arts syllabus itself was reorganized with eye to this new, lucrative demographic. Grammar, rhetoric and elementary logic (all with strong humanist flavors) were moved to the beginning of the degree: gentlemen could leave after a year or two, as many did, without a degree but with the solid grounding in the linguistic arts that a tutor used to provide, and which they needed for the law or civil service.[21] As dramatic as these changes were at the level of the university administration, they almost certainly lagged behind actual practice in the Colleges, which were providing an education ever more tailored to the needs of this clientele. The individual colleges grew in power and autonomy during this time, to the extent that a student's college would vouch that he had completed the course of study stipulated by the university and had attended the required lectures – whether or not he had actually done so.

It might be thought that such a change in the nature and very purpose of the arts degree might encourage a degree of intellectual philistinism – and that, indeed, is

[20] Curtis (1959, chapters 3 and 4).

[21] Feingold (1984, p. 30).

the impression we are given by an unlikely visitor to Oxford at this time. When Giordano Bruno paid his curious visit to Oxford in 1583, he found the scholars unreceptive to his lectures on Copernicus. The professors of Oxford suffered from two principal handicaps, he later declared: they were closed-minded Aristotelians, incapable of understanding the Copernican theory, and they were far more interested in speaking perfect, Ciceronian Latin than in knowing the truth.[22]

Largely on the basis of Bruno's report, Frances Yates argued that Elizabethan Oxford had broken entirely with its medieval past. The faculty remained devoted to Aristotle, but their devotion was to the pure, Greek text. They all but ignored the developments of more recent times, such as the remarkable kinematic theory of the fourteenth-century "Merton School." This puritanical fastidiousness about any departure from antiquity was just one effect of Oxford's embrace of fashionable Continental humanism. In their zeal for the new style of learning, Oxford scholars had rejected not only the dialectic of the sophisters, but also mathematics, astronomy and anything else that smacked of scholastic pedantry.

This was an intellectual change exacerbated by the political and religious turmoil of the previous half-century. Yates cited Anthony à Wood (writing in the late seventeenth century) who related that in 1550 there was a great purge of the Oxford libraries. Any "books wherein appeared Angles or Mathematical Diagrams" were burnt by government commissioners, suspecting that they were "Popish, or diabolical, or both." Merton was especially badly despoiled, "a cart load of MSS" being removed. The hole in the university's scholarly memory left only a "stiff, unyielding shell of dialectical habit," which the fashionable dons of Oxford filled (according to Yates and Bruno) with Greek and Latin prose composition.[23]

James McConica has argued against Yates's assessment of Oxford (and *a fortiori* Bruno's), through a careful examination of the Aristotelianism practiced in late sixteenth-century Oxford. He concluded that the intellectual culture of Oxford was not one of dry, sterile devotion to the text of Aristotle, but an eclectic, humanized Aristotelianism, centered around the practice of public debate and demonstration. While Oxford teachers had indeed abandoned the more rebarbative medieval dialectic, they had by no means forgotten their scholastic heritage.[24] Charles Schmitt showed that John Case (an Oxford contemporary of Savile's) combined humanist philological learning and eloquence with a deep knowledge of Aristotle and the Aristotelian tradition, to which he also brought the most recent speculation in the

[22] The 1584 *Cena de le ceneri* was Bruno's bitter lampoon of Oxford pedantry and philistinism; and in his *De la causa, principio e l'uno* written later the same year he laid out his side of the complaint against Oxford: see Bruno (1996, p. 81).

[23] Yates (1938–1939, especially pp. 230–231). It should be noted that evidence discovered after Yates's article showed that the *stated* reason for the cancellation of Bruno's lectures was not his Copernicanism, but his apparent plagiarism of a work of Marsilio Ficino's. See McNulty (1960), Aquilecchia (1963) and Aquilecchia (1995). It is still possible, of course, that Bruno was correct in his assessment of Oxford, even if he was not entirely candid about the circumstances of his dismissal – but this circumstance demonstrates at least that the Oxford dons were quite well-read in Renaissance Neoplatonism, as well as Aristotelian traditions.

[24] McConica (1979, especially pp. 298, 314–315).

Platonic and "hermetic" Continental thought; McConica concurred with Schmitt that Case could be seen as a forerunner of Francis Bacon.[25] McConica found little evidence of mathematical activity beyond the possession of astronomical books (some quite advanced) in private libraries; neither, however, did he discover any evidence of the fastidious distaste for the sciences that Yates claimed had overtaken Oxford.

There is, in fact, much evidence of mathematical activity at the universities – and not in opposition to the humanist culture of the age, but as a consequence of it. Paul Rose has shown that at Cambridge humanism and mathematics could be closely linked. The mathematical lectureship, held by scholars at Queens' and St. John's Colleges, was considered an element of the humanist revival at the university and was praised as such by Erasmus. Although its early occupants offered very elementary instruction (and could not plausibly be considered mathematicians themselves), it was later held by genuine practising mathematicians such as Thomas Hood and Henry Briggs; humanist mathematicians had "laid the institutional basis of mathematical studies at Cambridge."[26]

The fullest reply to both Curtis and Yates is Mordechai Feingold's book-length study on mathematics at the two universities.[27] Taking seriously Curtis's admonition that the university statutes did not adequately reflect the actual state of learning in the colleges,[28] Feingold uncovered evidence in manuscripts, letters, library records and annotated printed books of strong mathematical interests among Oxford and Cambridge students and teachers, supported by informal, yet often advanced mathematical instruction within the colleges. Mathematicians formed within the academy were able to hold their own with other European mathematicians. On their Continental tours, for instance, Henry Savile and his younger brother Thomas were able to collaborate as equals with the most accomplished European astronomers – a collaboration that was assisted and mediated by humanism, as they labored not only on mathematical calculation but also on the recovery of ancient astronomical and geometrical texts that might inform their work.[29]

Such a healthy picture of the state of the sciences at Oxford does not sit well with Bruno's complaints of humanistic pedantry. That is, perhaps, not surprising, considering the Italian philosopher's oversensitive and narcissistic nature. But his complaints were echoed by figures from within the university system – in particular, and most forcefully, by Henry Savile himself.[30] Savile complained throughout his early ordinary lectures that mathematics had all but disappeared from Oxford, despite the university's medieval supremacy in the field. At one point in his lectures,

---

[25] *Ibid.*, p. 310.

[26] Rose (1977, especially pp. 46, 58–59).

[27] Feingold (1984).

[28] Curtis (1959, p. 93).

[29] On Savile's tour, see Feingold (1984, 124–129); and for more detail of his astronomical work with fellow mathematicians, see Goulding (1995).

[30] See Goulding (2002), *passim*, for more details of Savile's complaints against the university, and the remarkably similar observations made by Henry Briggs in Cambridge 18 years later.

Savile connected the decline in mathematics quite explicitly with the recent vogue for humanist learning. Oxford, he told his students, in its pursuit of "eloquence, the Greek language and civic philosophy" was "now more Attic than Athens herself."[31] At one time, he went on, the whole world had looked to Oxford for instruction in mathematics, astronomy and physics; for many years, however, there had not been a single student with even an adequate grasp of these disciplines.[32]

Every element of the university, it seems, had to share the blame for this neglect of the sciences: the students themselves, who were too lazy to exert themselves in learning a subject they dismissed as useless and difficult; the teachers, who "day and night harp on at the same old erroneous ideas";[33] and the university itself, which in its statutes prescribed textbooks wholly inappropriate for the teaching of these subjects. Only remedy these faults, he concluded, and

> then indeed we shall have very many mathematicians. And not Oronce Fines as in France, nor Sebastian Münsters as in Germany, but Archimedes and Ptolemies, or rather Swinesheads, Bacons and Wallingfords, as there were in that long ago Oxford so different from our own; and we shall make this university, already famous through its profession of so many of the liberal arts, by far the most famous through the addition of great mathematicians.[34]

Savile had no grievance with humane learning itself; to the contrary, the early pages of his lectures were devoted to convincing his students that mathematics was actually *part* of the *studia humanitatis*.[35] And in later years he himself became renowned for his classical scholarship by translating the *Histories of Tacitus*, composing an account of the last days of Nero to bridge the gap between the *Annals* and the *Histories*, and editing the complete works of John Chrysostom, among many other lesser projects.[36] But to master mathematics, a special kind of humanism was needed, a delicate balance of philological sensitivity and scientific expertise (a balance which, not coincidentally, Savile himself possessed): "I believe that if I gave Archimedes to Cicero to translate, he could not do it without making frequent mistakes."[37] And, as he would reflect many years later in a speech before the Queen, modern Oxford's disdain for its mathematical past was not really founded on any philosophical objection, but was purely a matter of fastidiousness over the style of Latin used by medieval mathematicians and natural philosophers. Savile, though himself an accomplished humanist with an impeccable Latin style, could look beyond the

[31] MS Savile 29, fol. 3r: "... eloquentiae, graecae linguae, philosophiae civilis studia vigent apud nos tanta, ut ne ipsas quidem Athenas magis umquam Atticas extitisse putem."

[32] *Ibid.*

[33] *Ibid.*: "... eandem mendam diem noctemque tundentes..."

[34] *Ibid.*: "Nae permultos habebimus, non Orontios quales Gallia, non Munsteros quales Germania, sed Archimedes, Ptolemaeos, vel quales illud Oxonium huic nostro dissimillimum, Swinsetos, Bacones, Wallingfordos et Academiam per se ipsam tot iam disciplinarum professione claram, clarorum mathematicorum accessione longe clarissimam reddemus."

[35] On this theme in Savile's lectures, see Goulding (1999), *passim*.

[36] For details of Savile's humanistic scholarship, and bibliography, see Goulding (2004a,b).

[37] MS Savile 29, fol. 64r: "Et credo si Ciceroni dedissem Archimedem convertendum, non potuit non saepissime decipi."

language to the content:

> Since they[38] were completely provided with all the required gifts of both natural ability and learning, I do not mind in the slightest that they lacked elegance of style, in which we now glory almost exclusively, or at least excessively.[39]

But it would not be an easy task to return Oxford to its former mathematical glory; and Savile's more immediate problem was with the young men seated in his auditorium and their inadequate preparation for his astronomical instruction. Later in his lectures, expounding on the nature of arithmetic, Savile warned them of the rigors ahead and the grounding he would expect in this particular art. Those who could not count, or add and subtract simple numbers, he said, would be best advised to leave immediately.[40] He would also require a thorough knowledge of the first six and last three geometrical books of the *Elements* and recommended that they refresh their knowledge of these books in the evenings. He imagined some of his students complaining that Oronce Fine or Sacrobosco would not have required so much homework – as we shall see, these were elementary texts that Savile could have chosen to expound in his ordinary lectures. But so that no one would feel shortchanged, he promised to cover everything in these authors in a single lecture – "for (he said) it is completely absurd to waste an entire year in mere definitions and divisions of circles."[41] Somewhat disingenuously, Savile predicted that none of his students would have the slightest difficulty with the more advanced material he chose to present, "since they have profitably devoted three terms to arithmetic and two to geometry, not only attending ordinary lectures but also pursuing private study – they should have done so, and indeed I hope they have."[42] Savile returned to the customary teaching of astronomy at Oxford several times in these lectures. At one point he said that some might not even recognize the subject as he would teach it: "They think that the only kind of astronomy is the one that they themselves have learnt; and because they find nothing of this type in Sacrobosco or Oronce Fine, they consider that it is not astronomy at all, making their judgement not from reality but from their own laziness."[43] But real astronomy could not be learned from

[38] Earlier, he had referred to Roger Bacon, Walter Burley, Duns Scotus, William of Ockham and John Wycliffe.

[39] Plummer (1887, p. 265): "...quos, cum ab omnibus cum ingenii tum doctrinae subsidiis fuerint instructissimi, isto orationis flore, quo nunc fere solum, certe nimium, gloriamur."

[40] MS Savile 29, fol. 9r: "Nam neque conamur eum docere Ptolemaeum, qui numerare nesciat, nec speramus qui addere, subducere, multiplicare, dividere numeris huiusmodi non possit, eum aliquando planetarum epochous, eccentrotates, apogea perite numeraturum."

[41] *Ibid.*: "Quae tamen omnia, ne quis desit ad artis integritatem, una lectione comprehendi audietis. Perridiculum autem est, integrum annum in definitionibus et divisionibus circulorum, id est terminorum sola cognitione consumere."

[42] *Ibid.*: "Nec tamen, quamvis Sacroboscus vel Orontius forte non peraeque requirant, summam ideo rationem auditorum meorum non habeo, cum ii sint, aut esse debeant, et extitisse sperem, qui tres terminos Arithmeticae, duos geometricae, cum publicis in scholis, tum privatis meditationibus fructuose impenederint."

[43] MS Savile 29, fol. 13v: "Sed homines eam solam opinantur astronomiam, quam ipsi didicerunt, et quia nihil tale videant in Sacrobosco vel Orontio, nec esse quidem arbitrantur non ex rei natura, sed ex propria ignavia iudicantes."

simplified handbooks; it demanded the study of the greatest works in the genre, Ptolemy's *Almagest* and the writings of his modern follower (as Savile saw him), Copernicus.[44]

These were powerful charges against the mathematical culture of Oxford; we have to admire the sheer nerve of the young Savile turning the pedestrian genre of the ordinary lecture into a critique of the system of instruction itself. One cannot dismiss the possibility that there was some artful rhetoric behind Savile's complaints: by making the situation at the university seem so dire, his own accomplishment appeared all the more laudable. But whether things were as bad as Savile represented them, it is undeniable that Oxford had experienced changes through the Tudor period that had had an impact on the university's provision of teaching – and it is important to note that Savile's constant focus is on the failings of the *lecture system* and the university-level instruction in the sciences. Despite the liveliness of mathematical activity in the colleges (and Savile's own remarkable attainments provide evidence that it was possible to find a mathematical education outside of the "official channels") he clearly felt that the ordinary lecture system was failing.

By the late sixteenth century, students and teachers alike were entirely dissatisfied with the system of ordinary lectures – a fact evidenced not only by Savile's complaints but by the administration's repeated attempts to shore up the system in this period. In the early sixteenth century, a series of *de facto* reforms had been introduced, intended to stem the flow of defaulting regent masters. It became almost automatic to excuse regent masters from their duties as soon as a new class of MA graduates incepted, in effect reducing the period of regency from 2 years to one. Moreover, only a few masters were chosen to lecture in each subject; those who were excused paid a small amount of money (supposedly the fine for non-compliance, preserving a fiction of universal regency) which was passed directly as a stipend to the masters "actually regent" (as they were called), who delivered the ordinary lectures for that year. These reforms were made official and written into the university statutes in the 1550s.[45]

By Savile's time, then, ordinary lecturers had a degree of security: the small number of actually regent masters found plenty of students to teach and space in which to meet them in the Arts School, and they were guaranteed regular payment from the university out of the funds donated by non-teaching masters of arts. But the removal of competition for students seems to have had the opposite of the intended effect: lecturers no longer had any incentive to make the quadrivium interesting. In a university more and more devoted to humanistic studies, the ordinary lectures in the mathematical sciences fell into rapid decline. In the second half of the sixteenth century – despite frequent committees appointed to investigate the state of teaching – the university register continues to record defaulting lecturers and non-attending students, especially in the quadrivial arts. The music lecture fell almost entirely into disuse; lecturers were still appointed, but rarely fulfilled their duties. Matthew

[44] See passage cited below, at n. 77.

[45] Fletcher (1986, p. 186).

Gwinne, for instance, was appointed music lecturer at Oxford for the year 1582–1583. The university accepted his petition to be released from his duties because, as he put it, "this subject, if not entirely useless, is at least little practised."[46] The appointment of this lecturer had, in fact, become little more than a formality; the university register reveals that lecturers were automatically excused from teaching this subject, no further reason being needed than "students are not interested."[47]

The haphazard provision of lectures, particularly in the quadrivial arts, was bound to have the most profound effect on the sciences. A proper understanding of astronomy depended upon a good grounding in arithmetic and geometry. This was recognized in the statutes, which originally required all students to study arithmetic for one term and geometry for two, before they were permitted to attend the two terms of astronomy lectures; the statutes of 1564–1565 increased the arithmetic requirement to three terms.[48] This was a sensible arrangement, but one that was undermined by the habitual granting of exceptions to students, or even entire classes of students. The following petition was presented to Congregation in 1580:

> It is requested that those bachelors who are obliged to attend the geometry lecture be promoted to the astronomy lecture while the aforementioned lecture is intermitted. The reason for this is that Master Wignall, the public lecturer in geometry, has left town, called away on important business.[49]

The petition was granted.

It is clear that the sciences – at least as they were "officially" taught by the university – suffered during this period. Students were permitted to reach the higher levels of the science curriculum without a sufficient grounding in the essential preliminaries, and were taught in a stilted, out-moded manner. It was a system, moreover, which did not reward innovation or originality in its lecturers; there must have been powerful temptation to fulfil the university's requirements as minimally as possible, and to spend the year of regency expounding the simplest possible text.

## "An Ordinary Lecturer, That Is, Almost Less Than Nothing"

So Savile's appointment as ordinary lecturer in 1570 was no great prize; nor would much have been expected from his teaching. In his lectures, Savile said as much himself. Referring to himself as "an ordinary lecturer – that is, almost less than

---

[46] Clark (1887–1889, vol. 1, pp. 145–146): "Praxis eius scientiae si non inutilis at inusitata reputatur."

[47] *Ibid.*. Music students were often transferred to the "more useful" arithmetic lectures (Feingold 1984, p. 28, n. 14).

[48] Gibson (1931, p. 390).

[49] Clark (1887–1889, vol. 1, p. 99): "Supplicatur ut baccalaurei qui teneantur interesse lectioni geometriae promoveantur ad audiendam astronomiam pro tempore intermissae praedictae lectionis. Causa est quod Mr Wignall, publicus geometriae praelector, necessariis avocatus negotiis profectus est."

nothing,"[50] he admitted to some reluctance to take on the task of lecturing.[51] But, he went on, he intended to use this unglamorous position to revive the serious study of mathematics at Oxford. His ambitious plan started with his choice of a difficult text to expound: Ptolemy's *Almagest*.

In itself, this was an unusual choice of text, as can be seen from examining university statutes and contemporary practice. According to the statutes that had been in force since 1431, astronomy masters were expected to spend two terms expounding either the *Theorica planetarum* (an elementary textbook of planetary astronomy) or Ptolemy's *Almagest*.[52] But the inclusion of the *Almagest* should not be taken as an indicator of what was typically taught by the ordinary lecturers. If the teaching of the *Almagest* were common practice, one might expect to see commentaries on the *Almagest* prescribed in the statutes, or other works of similar mathematical sophistication (most obviously, Copernicus's *De revolutionibus*). But no such provisions appeared. Far from stressing the importance of teaching Ptolemy, successive statutes tended to recommend more *elementary* astronomical instruction, no doubt reflecting the actual state of mathematical teaching. Given a choice among several possible texts, the master appointed to the astronomical lectureship had no incentive to teach the more difficult text.

The Edwardian reforms of 1549 allowed the astronomy professor to lecture on geography instead of astronomy, using Pomponius Mela, the Elder Pliny, Strabo or Ptolemy's *Cosmographia*, a development which does not bespeak a great commitment to the study of the stars.[53] The statutes closest to Savile's time and in force when he delivered his lectures – the *Nova Statuta* of 1564–1565 – reaffirmed the previous regulations, while offering an even wider range of texts from which ordinary lecturers could choose. Lecturers in astronomy were now also permitted to expound the elementary *Sphere* of Sacrobosco. In admitting this text, the university was almost certainly only giving official sanction to an already established practice: throughout Europe, Sacrobosco's basic astronomy handbook was usually taught as an introduction to the planetary *Theorica*, long prescribed by Oxford statute as an approved astronomical text.[54] Another decree, issued later in 1565, clarified the teaching requirements and made recommendations of specific texts that were most suitable for ordinary lectures. Astronomy lecturers, according to this document, were especially encouraged to expound either Sacrobosco's *Sphere* or another

---

[50] MS Savile 29, fol. 3v: "praelector ordinarius, id est paene minus quam nihil."

[51] *Ibid.*, fol. 2v: "Suscepto professionis istius onere, sponte an secus nihil ad hoc tempus..." ("Having taken on the burden of this teaching – willingly or otherwise, at the moment it matters not...").

[52] Statues of 1431, in Gibson (1931, p. 234).

[53] Statutes of 1549, in Gibson (1931, p. 344): "Mathematices professor, si cosmographiam docet, Melam, Plinium, Strabonem aut Ptolomeum enarret."

[54] Statutes of 1564–1565, in Gibson (1931, p. 378). On Sacrobosco's *Sphere* and its influence, see Thorndike (1949). For the close connection between Sacrobosco and the *Theorica*, see Pedersen (1981, pp. 114–115).

elementary textbook on spherical astronomy, written by the French mathematician Oronce Fine; the *Almagest* is not mentioned.[55]

Savile had some discretion, then, in selecting the astronomical text to be read in his lectures. His choice, the *Almagest*, was the most difficult text permitted by the statutes, and the only one of any mathematical sophistication. In his first lecture, Savile himself reflected on the fact that he was behaving contrary to expectations. Comparing his intention to teach astronomy to the beginning of a long sea voyage, he wrote:

> If, in teaching this art, I rehearse the monstrous hypotheses of the Alphonsines (or rather the stories of raving old men), then I shall necessarily be stuck in the shallows – yet the rehearsal of them is decreed. Good God, what intense ill-will I shall have to suffer, if I express my true opinion on the whole of mathematics![56]

The object of Savile's derision was the medieval *Theorica planetarum* which, as we have seen, was *not* a compulsory part of the astronomy course, as Savile represented it here.[57] But it appears to have been the expected choice for the astronomy professor. Later in the lectures, Savile railed against professors who watered down the teaching of astronomy to the exposition of a mere handbook. For Savile, the term "astronomy" meant

> that art which demonstrates the forward and retrograde motions and revolutions of the planets and fixed stars. *I do not mean that discipline which is the only one most people call astronomy, consisting of the drawing of circles and illustrating by examples.* When I see this practice, I am so far from the contemplation of divine providence, so far from wonder at divine workmanship, that I marvel only at the shameless vanity of those who are so delighted with acorns when fruit is available.[58]

In the emphasized passage, Savile claimed that most people thought of astronomy as nothing more than the "drawing of circles," a fair description of the largely qualitative approach of the *Theorica*. It is worth keeping in mind that Savile himself had

---

55 Gibson (1931, pp. 389–390): "hos potissimum ad explicandum adhibento ... Orontium de Sphaera vel Iohannem de Sacrobosco in astronomia."

56 MS Savile 29, fol. 2v: "Monstruosas Alfonsinorum dicam hypotheses, an delirantium senum fabulas ad artis praescriptionem si revocaro, haeream in vado necesse est. et tamen revocare certum est. Si liberam de mathematicis omnibus dixero sententiam, quantus, dii boni, subeundus ardor invidiae?"

57 His dismissal of the hypotheses as the "stories of raving old men" was clearly intended to recall Johannes Regiomontanus's furious attack on the *Theorica* published in 1476, entitled *Disputationes contra Cremonensia in planetarum theoricas deliramenta*. The *Theorica* was very occasionally attributed to Alfonso of Castille, the great patron of astronomy, with whom it had no connection. Nor, *pace* Regiomontanus, did it most likely have anything to do with Gerard of Cremona. See Pedersen (1981).

58 MS Savile 29, fol. 20v: "Astrologiam autem eam intelligo, quae progressus, regressus, conversiones luminum errantium fixarum[que] demonstrat, *non eam quam plerique solam esse opinantur, qua circuli definiuntur, describuntur, exemplis illustrantur,* quae cum video, tantum absum ab ea quam extuli providentiae cogitatione, tantum ab admiratione fabricae divinae, ut nihil pene admirer, quam eorum impudentem vanitatem, qui frugibus inventis tantopere glandibus delectarentur." Emphasis mine in Latin and translation.

sat through ordinary lectures on astronomy only 2 or 3 years earlier, as an MA student. It is surely significant that, as an ordinary lecturer reflecting on his formation as a mathematician, he never once mentions the influence of the ordinary lecturer whose astronomy lectures he himself attended. It is not implausible to suppose that his angry denouncements of mathematics lecturers and their inadequacies are based on bitter experience as a student. Still, was this a fair description of the usual practice of other lecturers? Savile's is the only complete set of lectures at Oxford that survives from this period;[59] there is no other record of the texts that ordinary lecturers in Savile's day actually expounded. So we have no proof that the situation was as bad as he claimed. However, the records of a slightly later lecturer do seem to corroborate Savile's complaints, confirming that the statutes could be read as permitting astronomical instruction at a very elementary level.

In 1591, Francis Mason of Merton College delivered the ordinary lectures in astronomy. His English notes for the lectures survive.[60] Over some twenty pages, Mason made an elementary *précis* of Sacrobosco's *Sphere* and the medieval *Theorica planetarum*. He wrote out a detailed explanation of the system of concentric spheres for the motion of the sun and moon, and made careful drawings of the circles – without, however, explaining why these particular circles were required. Nor does anything remotely mathematical appear in his notes. Mason did not even write down the numerical parameters for the spheres of the luminaries (their relative radii and rates of motion), which were almost the only numerical data included by the author of the *Theorica*. In an age in which Copernicus's name had become commonplace, Mason said nothing of the new astronomy – in fact, he mentioned no other authors at all, never looking beyond the texts set before him to expound. In other words, Mason gave his students nothing more than a general, qualitative (and outdated) sense of the arrangement of circles in the heavens. This could hardly be called instruction in astronomy, and was, no doubt, the sort of thing that Savile had in mind when he decried the "sterile drawing of circles" in his ordinary lectures some 20 years before. Such an approach clearly remained popular (or, at least, convenient for teachers), despite Savile's efforts to educate his contemporaries.

Without doubting the sincerity of Savile's complaint that the study of mathematics had declined at Oxford, nor his analysis of the reasons, nevertheless it is true that his lament also rang the changes on a humanist commonplace: the status of learning and its deterioration since classical times. Humanist mathematicians, in particular, made a habit of bewailing the state of their discipline, as mathematics not only shared in the general misfortune that had befallen all the arts of classical antiquity, but also suffered the additional burden of a reputation for difficulty and uselessness. This topos is found, for instance, in the preface to Regiomontanus's *Epitome in Almagestum*, a work that Savile certainly knew. Regiomontanus wrote that the inventors of the liberal arts were virtuous men, unconcerned with financial

[59] John Chamber's 1575 lecture notes (see Conclusion, at p. 180) were largely copied from Savile's. On the Cambridge lecture notes of Henry Briggs, see Goulding (2002).

[60] MS London, British Library, Harley 6494, fols 57–77.

gain, and dedicated to their art for its own sake. The inevitable decline from this Golden Age set in when a desire for acquisition began to creep into men's minds. The eventual result was the terrible state of mathematical learning that Regiomontanus said he found in his own era. Why then did no one make the effort to revive these arts? Mathematics, Regiomontanus explained, was commonly perceived to be very difficult, and so scholars were unwilling to devote themselves to it. This was due to the intrinsic complexity of the subject matter and the poor state of the books, which in turn was the fault of translators incompetent both in classical languages and the sciences. In their *Epitome*, Regiomontanus and his teacher Georg Peurbach had not only produced a better version of the Almagest, he claimed, but had also given some thought to order and presentation, so as to make the work more suitable for the student.[61]

In short, Savile may have been right to criticize Oxford, but in doing so he was also setting himself up publicly to look like the kind of person who *could* criticize the contemporary scene: someone with the status of a Regiomontanus, or even a Ramus. As we turn now to the content of the lectures themselves, we will find that they really only make sense when read against these various contexts: Savile's philological humanism, his Platonism, his concern for institutional reform at Oxford, his abiding interest in the mathematical accomplishments of his colleagues on the Continent, and the formative (if one-sided) debate he carried on with Ramus over the proper status of all of these in the modern academy.

## The 1570 Lectures

Savile wrote out the text of his lectures in three volumes, now preserved in the Bodleian Library.[62] The lectures divide naturally into four sections:

1. Protreptic exordium (MS Savile 29, fols 2r–8r).
2. Introduction to the seven mathematical sciences (according to the classification of Geminus): arithmetic, geometry, music, optics, mechanics, astronomy and geography (MS Savile 29, fols 8r–25r).[63]

---

[61] Regiomontanus (1550, sigs A2r–A3r). Many of the opinions Savile expressed in his protreptic on the decline of mathematics and the measures needed to restore it to its proper place echoed those of yet another mathematical humanist, Francesco Maurolyco. Maurolyco's thoughts were stated in manuscript works which Savile could not have seen; this does, however, illustrate how men who were educated in both the philological tradition of humanism and the mathematics of antiquity tended independently to very similar conclusions. See Rose (1975, ch. 8). The similarity to Ramus's model of decline is also apparent.

[62] MSS Savile 29, 31 and 32. MS Savile 30 contains John Chamber's lecture notes (see p. 180).

[63] Geminus's division of the sciences was recorded by Proclus in his *Commentary* on Euclid. See Proclus (1992, pp. 31–35). In his copy of the *Commentary* Savile made many annotations to this section, summarizing Geminus's division in the margin (Savile W.9, p. 11). He defended this division, over the more usual quadrivial division into arithmetic, geometry, music and astronomy, at MS Savile 29, fol. 8r.

3. History of mathematics from Adam to Ptolemy (MS Savile 29, fols 29r–65v).
4. Commentary on the mathematics of the *Almagest* (remainder of MS Savile 29, fols 65v–140r, and MSS Savile 31 and 32), presenting also Arabic and Copernican astronomy alongside the Ptolemaic models.

We have already seen something of the protreptic section and the account of the seven mathematical arts, with which the lectures open. But the central portion, perhaps the heart of the entire project, was his historical excursus. In many ways the excursus stands apart from the rest of the lectures – in no small part because it was envisioned as a separate work in the first place. How Savile came by the information in the excursus helps set the stage for a sustained reading of all three introductory sections of the lectures.

In the late 1560s, as he was teaching himself astronomy by translating the *Almagest*, Savile had also begun to take an interest in the history of mathematics. In the blank pages at the back of one of the notebooks of his translation, he compiled an impressive bio-bibliographical list of *auctores mathematici*, research that he would draw on for the historical section of his 1570 lectures.[64] Savile listed nearly seven hundred writers on mathematics, astronomy and the other sciences. He included ancient, medieval (both western and Arabic) and contemporary authors.[65] For some authors, Savile had nothing more than a name; but generally he provided brief bibliographical information and a list of the author's published and, where possible, unpublished works – even the location of their manuscripts.

To compile this enormous repertorium, he used, for the ancients, Diogenes Laertius's *Lives of the Philosophers* augmented by other ancient and humanistic histories; and for modern authors and bibliographical information about the ancients, he consulted the massive bibliographies of Konrad Gesner and John Bale. To fix the dates of many of his *auctores* he plundered the chronological sections of Caspar Peucer's astronomical handbook, the *Elementa*. But one text in particular stands out as the source for dozens of his entries: Ramus's *Prooemium mathematicum* in its first edition of 1567.[66] Savile drew almost all of his information about contemporary European scientists from the *Prooemium*, while many of the entries on

---

64 The list begins on the page following his translation of book V of the *Almagest* in MS Savile 28. He thus must have drawn up the list after he had translated this part of the *Almagest*, and before the 1570 lectures, in which he puts to work his research on mathematical authors – that is, between late 1568 and October 1570. The translation ends on fol. 28r; the list of authors begins on fol. 28v. After starting the list, Savile renumbered fols 29 to the end of the manuscript starting from 1. Here I shall refer to what originally was, say, fol. 30, as fol. *2, using the new numbering

65 The list is arranged in several roughly alphabetical sequences. It seems that Savile compiled the list over some time, making notes on mathematical authors (in papers that have not survived), which he periodically copied out into MS Savile 28 in alphabetical order, after which he continued to compile new authors.

66 On the rear flyleaf of MS Savile 28, Savile wrote "Gesneri bibliotheca, Balaei centuria, Diogenes Laertius, Peuceri Sphaerum, Prooemium Mathematicum," and scribbled Ramus's name several times elsewhere on the page. None of Savile's copies of these sources have survived (except, perhaps, the Peucer, a copy of which is in the Savile collection of the Bodleian Library (shelfmark Savile Aa. 14; the copy is without annotations or markings)).

ancient scientists were built around summaries of Ramus supplemented by his other research sources. Frequently, Savile gave a page reference to the 1567 *Prooemium*;[67] in some cases, the entire entry consisted of a reference to Ramus.[68] To give an example of how he wove together his sources, his entry on the ancient arithmetician Diophantus (the second author in the entire list) will suffice:

> Diophantus, on polygonal numbers. Two books of arithmetic with scholia by Maximus Planudes and another unnamed writer. The Greek is found in Rome and elsewhere in Italy. Ramus: we have six books in Greek, even though the author promises 13. He is cited by Theon.[69]

Savile copied the first part of the entry from Gesner's *Bibliotheca* – he omitted only the bibliographer's notices of a work by the same author on music, a subject in which Savile had little interest.[70] The rest of the entry he took, as he said, from Ramus's *Prooemium*, focusing on the information that Ramus gave there about the number of books of the *Arithmetica* (about which Gesner was uncertain) and the location of the manuscripts.[71]

In summary, the manuscript evidence shows that while Savile was finishing his *Almagest* translation late in 1568, he studied Ramus's *Prooemium* and began to use it and other reference works to draw up his list of *auctores mathematici*. It was also around this time that he began to consider writing a history of mathematics of his own to rival Ramus's *Prooemium*. On one of the blank pages of the first volume of his *Almagest* translation, Savile headed a page with the title "Compendium historiae mathematicae;" he later crossed out the title, however, and nothing more on the history of mathematics appears in the volume.[72] Perhaps Savile realized that a few blank pages at the end of a notebook would not be sufficient for his thoughts on the history of his discipline. He took the opportunity to expand, at great length, on his research into the history of the mathematical arts, when the following year he was chosen by the university to deliver the ordinary lectures in astronomy.

---

[67] In his quite extensive entry on Proclus (at fol. *13r), for example, he gave a comprehensive list of Proclus's mathematical works and a little biographical color. Some of the bibliographical information he drew from editions of Proclus's work in his own library; but most of the information (particularly the philosopher's biography, his worth as a mathematician and the existence of other ancient Procluses) he took from pp. 154–155 of the *Prooemium* (p. 37 of the *Scholae mathematicae*), for which he cited the page references.

[68] For instance, his entries for Caspar Peucer and Conrad Dasypodius, both on fol. *6v, read in their entirety "281 Rami" and "284 Rami" respectively.

[69] MS Savile 28, fol. 28v: "Diophantus. de numeris polygoniis. item lib. 2 Arith. cum scholiis max. planudis, et alteri innominati. graece servantur Romae, et alibi in Italia. Ramus: 6 libros cum tamen author 13 polliceatur graecos habemus. citatus a Theone."

[70] Gesner's entry (Gesner, 1545, fol. 214r) reads: "Diophanti scriptoris Graeci arithmetices libri duo (alias, sex) cum scholiis Max. Planudis et alterius innominati. Harmonica, et quaedam de numeris polygoniis. Omnia Graece servantur Romae et alibi in Italia."

[71] Ramus *Prooemium mathematicum*, p. 120 (= p. 37 of the *Scholae mathematicae*): "Diophantus, cujus sex libros, cum tamen author ipse tredecim polliceatur, graecos habemus de arithmeticis admirandae subtilitatis artem complexis, quae vulgo Algebra arabico nomine appellatur; cum tamen ex authore hoc antiquo (citatur enim a Theone) antiquitas artis appareat."

[72] MS Savile 26, fol. 81v.

And so, just as we have read Ramus's history of mathematics within a certain context – the fallout of the bitter dispute with Charpentier over the chair in mathematics at the Collège – so too must we consider the purposes for which Savile wrote his history. His historical researches may have started life as a collection of *auctores mathematici* and have been intended for a *Compendium* on the history of mathematics; but, in the form he made it public, it was addressed to Oxford students and must be read as a piece of rhetoric directed specifically to this audience.

## The Place of History in Savile's Lectures

The students who attended Savile's first lecture on October 10, 1570, had no idea how ambitious his lectures on the *Almagest* would be.[73] The passionate rhetoric of their lecturer's exordium may have given them some clue, however, that his ambitions ran far beyond imparting the elements of the "sphere" or the use of an almanac:

> When Aristippus, the Socratic philosopher, was shipwrecked on the shore of Rhodes, he found that his small band of companions in that same ill fortune and peril were greatly afraid: some feared that, although they had survived the waves and had thought that they were out of danger, they would now starve to death in a forsaken land; others, having braved the rocky, barbarous sea, trembled at the thought of beasts more terrible than any storm; and still others shuddered with fear of meeting men more dangerous than any beasts, human only in appearance. But Aristippus saw, drawn in the sand – that sand which can never be praised enough! – some mathematical diagrams, and took it as a very great sign of hope for them all. "The greatest dangers and wildest storms are now past," he said. "Look in the sand, my friends, and see the calculations. See the circles, triangles, squares, polygons; in my misery the contemplation of them delights me, raises my spirits in my depression and consoles me in my suffering. I can tell you – not as an augur from the birds, not as a soothsayer from entrails, not from the stars as an astrologer (and never have I regretted my ignorance of *their* occult mysteries) – but as, perhaps, a prudent judge of our situation, I can prophesy from these drawings in the sand your safety and the end of all your miseries. These figures are tokens of humanity, and are no small mark of Greek learning. Believe me, my friends, the study of these arts is incompatible with a savage mind; these arts are noble and are learnt by noble men. Nor can anyone embrace the liberal arts unless he is liberally educated. Have high hopes for the character of these islands; those who know how to "geometrize," know how to show mercy.[74]

---

[73] On the difficult question of the length of Savile's series of lectures, see Appendix B.

[74] MS Savile 29, fol. 2r: "Socraticus Aristippus, cum ex naufragio Rhodiorum ad littus proiiceretur non multis comitatus eiusdem periculi fortunaeque sociis, pertimescentibus caeteris, partim ne sibi qui superatis iam fluctibus omni se molestia defunctos arbitrabantur, nova necessitas instaret in agro deserto fame pereundi, partim ne pelago iam usi scopuloso atque barbaro, bestiis uterentur deinceps aestu quovis immanioribus, et partim ne in homines inciderent belluis infestiores, nihil humani praeter faciem habentes; primus conspectis mathematicorum diagrammatis et illo numquam satis laudato pulvere ad bene sperandum de salute omnium quasi signum aliquod amplissimum extulit. Maximas molestiarum moles et turbulentissimas tempestates effugimus. En illum, comites, pulverem et abacum. En circumductos circulos, descripta trigona, tetragona, polygona, quorum me contemplatio maerentem delectat, iacentem erigit, afflictum excitat. Non ego

The echoes of the Aristippus of Melanchthon's preface (which Savile surely had in front of him as he wrote this) are unmistakable.[75] And like the German reformer, Savile drew from Vitruvius's story the message that mathematical ability implied a degree of moral development. He also followed Melanchthon by contrasting Aristippus's relief on finding geometrical drawings in the sand with his own disappointment at the mathematical attainments of his countrymen. The story of the shipwreck was the principal metaphor of Savile's first lecture, in which he compared his decision to lecture on the *Almagest* to a dangerous sea-journey, forever at risk of being wrecked among those who cared nothing for mathematics whatever.

By beginning his lectures in this way, Savile indicated that he would approach mathematics in a characteristically Melanchthonian way: mathematics was to be seen as one of the liberal arts, with essentially Platonic ends. (It is impossible to tell whether his students were astute or well-read enough to *recognize* the significance of this anecdote). His Aristippus, like Melanchthon's, served to guarantee that mathematics had the highest of all goals – however much its critics might pretend to disdain it.

Throughout the protreptic, Savile continued to deploy historical anecdotes in order to recommend mathematics to his students. Much like any writer of Renaissance epideictic oratory, Savile used historical *exempla* to underline his moral points.[76] This was a rhetorical strategy familiar to his students from their humanistic studies elsewhere in the syllabus. Furthermore, by opening his course with so strong a declaration of humanistic values, Savile reassured his students that they would not be studying from scholastic handbooks of "spheres," but drawing from the very founts of antiquity itself. In his opening lectures, he even framed Copernicus's new astronomy as a historical artefact:

> But someone will say, did Copernicus not add something to astronomy? And what about the countless other books written about astronomy? All the others I will not hesitate to reject out of hand. For what is in all these little books on the sphere that is not already treated much more abundantly and clearly in Ptolemy? As for Copernicus, he has indeed earned immortal fame; but he has not added anything new to astronomy that was not already thoroughly discussed by Ptolemy. Instead, he has clarified the same problems by means of a new method, with different hypotheses.[77]

---

vobis, ut augur ab avibus, non ut aruspex ab extis, non a stellis, ut astrologus, quorum occultis mysteriis carere me non moleste fero, sed ab hisce depictis formulis, ut non imprudens forte rerum aestimator, salutem et miseriarum omnium finem denuntio. Figurae sunt humanitatis indices, graecae disciplinae non leve vestigium. Mihi credite, comites, harum artium studia in animum agrestem non cadunt. Ingenuae sunt, ab ingenuis discuntur, nec quisquam, nisi liberaliter institutus, liberales artes complecti potest. De moribus insularum bene sperate. Sciunt misereri qui sciant *geômetrein*."

[75] See passage cited at n. 61 of first chapter.

[76] See, for instance, Trinkaus (1960); Vickers (1988, pp. 744–745); and Plett (2004, p. 146).

[77] MS Savile 29, fol. 23r: "Quid igitur, inquiet aliquis, nihilne adiecit Copernicus, nihil tot de astronomia perscripti libri? Caeteros quidem omnes non dubitabo mea sententia condemnare. Quid enim tot sphaericis libellis continetur, quod non extet apud Ptolemaeum multo uberius, multo illustrius? Copernicus, quem laudes immortales meruisse constat, non aliquod novum caput ad

Savile's students had no doubt heard of Copernicus and his famously challenging *De revolutionibus*. Savile glossed over the difficult technical innovations they would face when he came to teach these new models, and emphasized rather the continuity between the new astronomy and that of the ancient world. Later, in his history of mathematics, Savile mentioned Copernicus in the context of his biography of the ancient heliocentrist Aristarchus. Again, Savile downplayed any idea that the modern astronomer was an innovator, here going so far as to suggest that Copernicus's work was primarily of antiquarian, philological interest: the writings on planetary astronomy by Aristarchus himself had been lost, but from Archimedes's brief descriptions of the ancient astronomer's system and Copernicus's more recent work, the loss could be repaired.[78] Savile treated Erasmus Reinhold in a similar way. The great German astronomer had labored above all else on reforming astronomical tables on the basis of the new Copernican models.[79] But, for Savile, Reinhold was a man both "born to advance mathematics and deeply read in Greek literature;" he praised him particularly for his insights into the text of the *Almagest*.[80] Savile's treatment of Reinhold illustrates very clearly his own ideals and priorities: while he held the highest aspirations for mathematics at Oxford, he framed the innovations he wished to introduce as a return to an authentically ancient past.

Savile had been chosen as an ordinary lecturer in astronomy and, after his introductory, protreptic lecture, he might have been expected to start on the explication of Ptolemy and the teaching of astronomy. Instead, he turned to a meticulous description of each of the mathematical sciences (following a division attributed to the ancient mathematician Geminus), devoting several pages (the equivalent of as many as three entire lectures) to surveys of arithmetic, geometry and the other arts.[81] Deep in his discussion of mechanics (and some eighteen pages into his digression on the sciences), Savile wondered whether some might be growing impatient with his prolixity. Beyond simply wishing to say something about *all* the mathematical arts,

---

astronomiam adiecit, quod non esset a Ptolemaeo pertractatum, sed ipse easdem res nova quadam ratione variatis hypothesibus illustravit.

[78] MS Savile 46v–47r: "... crederem profecto, si pythagorica *metempsuchôsis* mihi probaretur, animum Aristarchi multa secula vagantem in corpus commigrasse Copernici. ... et quamvis ex suis ipse scriptis Aristarchus non potest cognosci ... eadem fere dicit [Archimedes] de astrologia Aristarchi quae sunt a Copernico nuper in caelo confirmata. itaque caelum hoc copernicianum novum quoddam inventum non est, cum quadringentis ante Ptolemaeum annis sit ab ingeniosissimo artifice constabilitum." ("If I accepted the Pythagorean doctrine of reincarnation, I should believe that the soul of Aristarchus, having wandered many centuries, had migrated to the body of Copernicus. ... And although it is not [now] possible to read Aristarchus's writings, ... Archimedes says almost the same things concerning the astronomy of Aristarchus as were recently affirmed in the heavens by Copernicus. And so this Copernican heaven is not some new invention, since it was established by a brilliant master four hundred years before Ptolemy.")

[79] See Gingerich (1973).

[80] MS Savile 29, fol. 64r: "Erasmus Rheynholdus, vir ad amplificanda mathemata natus, et graecis libris eruditus."

[81] See n. 63 above.

not just astronomy, he explained:

> I do, in fact, have good reasons for talking about all of these, because our history will embrace *all* mathematicians, and not just astronomers. And not, moreover, in just a historical fashion – what age they lived in, what manner of life they led, what country they inhabited – but rather mathematically: what they wrote in what field, how well they wrote it and how useful it is for teaching beginners. Since I intended to say this, I could not, without fault, omit a discussion of the whole of mathematics and each of its branches. For, if I were to say that Archimedes wrote eruditely on equilibria, Ptolemy on catoptrics, [Michael] Stifel on surds and irrational numbers,[82] [Girolamo] Cardano on algebra,[83] someone else on geodesia or another on astronomical fractions, without having first explained the power of each art, and with most [of you] still ignorant of its capacity, then that would indeed seem senseless and mad.[84]

This was the first indication that Savile intended to lecture on the history of mathematics. His students had already heard many historical anecdotes in the protreptic and in Savile's introduction to the individual arts; now Savile revealed his intention to lecture on the entire history of mathematics, before he had even started on the required exposition of the *Almagest*. As this passage shows, his plan was to cover not only the history of ancient mathematics – including Ptolemy and Archimedes – but also the developments of quite recent times (Stifel had died only 3 years earlier, and his major publication – the *Arithmetica integra*, had appeared a quarter-century before). In other words, Savile intended to cover much the same ground as his model and rival Peter Ramus: having defended mathematics from charges of difficulty and obscurity and having urged his audience to its pursuit, now Savile, like Ramus, would trace the history of the discipline from its very origins until the present day. The goals he had for his history, moreover, were much the same as those Ramus had pursued in his *Prooemium mathematicum*: to unfold the history of mathematics, to appraise the work of his predecessors, and to make a larger point about the nature and purpose of the discipline as a whole.

As Savile formulated the task, however, he intended to proceed more systematically than Ramus. Savile made an important distinction between presenting the biographies of mathematicians "historically," and assessing them "mathematically." Ramus, of course, also wished to weigh up each mathematician, paying special attention to their accessibility for beginners (a criterion that Savile embraced as

[82] Savile was referring to Michael Stifel's *Arithmetica integra* (Nuremberg 1544). Savile's unannotated copy of this edition is in the Bodleian Library with shelfmark Savile R.13.

[83] Here Savile almost certainly intended Cardano's *Ars magna sive de regulis algebraicis* (Nuremberg 1545). Savile's sparsely annotated copy of this edition is found at Savile N.15(2).

[84] MS Savile 29, fols 17r–v: "Quae tamen a me certis de causis non afferrentur, nisi cum historia nostra mathematicos omnes non solum astrologos complexura sit, nec historico more tantum, quo seculo vixerint, quibus moribus extiterint, quo caelo usi sint, sed mathematice magis, quid in quo genere scripserint, quam bene, quam ad instituendos tyrones commoditate. Haec, inquam, cum esse[n]t dicenda disceptationem et de tota mathesi et de singulis formis absque scelere non potui praeterire. Etenim si dicerem erudite scripsisse de isorrhopicis Archimedem, Ptolemaeum de catoptricis, Stifellium de surdis et irrationalibus numeris, de cossicis Cardanum, alium de geodesia, de scrupulis astronomicis alium, non explicata prius artis cuiusque vi et facultate iam plurimis incognita, amens profecto et insanum videatur."

well); but he hardly separated his critical role from the more empirical task of establishing the names, dates and accomplishments of his protagonists. Savile's history was also colored by his own pedagogical and philosophical concerns – still, he did make more of an effort to report the "plain (mathematical, as well as historical) facts" before adding his editorial observations.[85]

At the end of his lecture on the seventh mathematical art (geography), Savile promised his audience that he would begin the following day with an exposition of Ptolemy.[86] Overnight, however, he seemed to have had a change of heart; for in the next lecture he instead embarked upon his history of astronomy and mathematics, beginning with Adam. This was no mere digression; Savile's lectures on the history of mathematics occupy 68 closely-written pages of his manuscript (some 40,000 words), in which he recounted the lives and works of 154 astronomers and mathematicians. By my count this would have taken him at least 5 days to deliver, with each lecture several hours long.

Perhaps Savile really did decide at the last moment to make use of the extensive research he had pursued in the history of mathematics, and to insert an account of the science into his lectures. But he had already (perhaps inadvertently) signalled to his students that he intended to share "historia nostra" with them. It is much more likely that Savile had always intended to depart from the expected format of an ordinary lecture; his profession that he intended to turn to the *Almagest* was an artful way to excuse the roundabout manner in which he was expounding a standard text: a thoroughly unorthodox approach for the time, as we have seen, but no doubt very welcome to his students.

Nevertheless, Savile seems to have been genuinely wrong-footed by the *scale* of his history, once he set about telling it. He had accumulated so much information on ancient mathematics and mathematicians that his narrative threatened to occupy his entire course of lectures. Savile had told his students that his history would extend to modern times. That was the rationale behind presenting long introductions to *all* the mathematical arts, not just astronomy. Such a history would rival Ramus's; and would give Savile the opportunity to use the notes he had compiled on contemporary and medieval, as well as ancient mathematicians in his collection of *auctores mathematici*. But by the time Savile reached the life of Ptolemy, he worried aloud whether he would ever complete the exposition of his text within the allotted time. On an additional page at the very beginning of his notebook he added a paragraph to go at the end of Ptolemy's biography in order to ease the transition from his history

---

[85] The nuances to the term "history" in the sixteenth century were manifold, as a recent collection of papers has demonstrated (Pomata et al. 2005). See, in particular, Pomata (2005, especially p. 106–114). Savile may have had in mind a common Aristotelian sense of "historia" as "knowledge without causes" – a kind of bare narration of facts, which did not attempt to make an assessment from the point of view of any more specialized *ars*. Savile's statement that he would proceed *mathematically* as well as historically seems to be meant to assure his audience that he will enter into causes from a mathematical point of view – while yet keeping the two approaches distinct.

[86] MS Savile 29, fol. 25r.

(which, together with the introductory material that preceded it, he now dismissed as "talkative mathematics,") to his detailed exposition of the *Almagest*.[87] Once his attention turned to the actual text chosen for the ordinary lectures, he moved very quickly through the early chapters of Ptolemy; he had already, in his own protreptic, expatiated at length on the philosophical material in *Almagest* I.1, and perhaps at this point he had little patience for Ptolemy's own "talkative mathematics" in *Almagest* I.1–9. From chapter I.10 (which marks the beginning of Ptolemy's mathematical astronomy), Savile proceeded at a careful pace through the whole of the *Almagest*, introducing Arabic and medieval astronomy, and the new astronomy of Copernicus as elucidations of Ptolemy's great work.

## The Origins of Mathematics

The first few pages of Savile's history are concerned with the very origins of mathematics. As he did throughout his history, Savile divided up his narrative into sections, each focusing on a few principal protagonists whom he listed at the beginning of each section. In the first sections, Savile focused in turn on "the sons of Seth," "Noah," "Samothes and the Druids," "Zoroaster and the Magi," "the Chaldeans" and, to bring him into the age of the Hebrew Patriarchs, "Abraham, the Phoenicians and the Egytians."[88] From even this brief list of titles, the broad outlines are familiar enough; as ever, it is clear that Savile depended on Ramus above all other sources. All of the familiar elements were present: astronomical activity in the Garden of Eden, the two pillars that preserved mathematics from the Flood, Abraham's role as a "planter of mathematics," and so forth. Savile even parroted Ramus's oft-repeated dictum that arithmetic and geometry must necessarily have preceded astronomy, even if there were few sources that documented their first appearance. (Perhaps aware that this was statement was too obviously Ramist, Savile cancelled out the entire sentence in his manuscript.)[89]

However much Savile owed to Ramus for the basic outline of his history (and even many of the details), nevertheless he also showed a great deal of independence of thought throughout his lectures. He had first-hand familiarity with all of the sources that Ramus cited, and brought in many others that he himself had discovered. At the very beginning of the history, he praised Josephus as a source for the story of the sons of Seth, setting out some historiographical principles under which he accepted the ancient account. Josephus was an eye-witness (insofar as he claimed to have seen one of the antediluvian pillars himself); moreover, he had access to Jewish records that were now long lost, which he referred to by name. His account of Seth was corroborated by the *Suda*, in which the ancient patriarch was

---

[87] MS Savile 29, fol. 65r and the addition to this page on fol. 1v (*garrula mathesis*).

[88] MS Savile 29, fols 29r–31r. A complete account of the contents of Savile history may be found in Appendix A.

[89] *Ibid.*, fol. 29r.

said to have "named the stars." Savile observed that many others took Adam as the first and most perfect mathematician – he was, perhaps, thinking of Ramus, who had often listed Adam as the very first practitioner; but Savile offered no opinion on this claim, noting only that another *son* of Adam, Cain, was also credited with the invention of a branch of mathematics, that of weights and measures and simple surveying.[90]

In his section on Noah, Savile continued to follow Josephus (and Ramus) closely. But he made two notable departures. First, he utterly rejected the idea that Adam had used astrology to predict the coming Flood (not, in fact, stated by Josephus, even if he implied as much). Although he generally took Josephus as an "oraculum sanctissimum," he had to believe that Adam learned of this looming event from God, not by observing the stars. To believe otherwise would be to concur in the impious superstition of those who cast horoscopes for the Lord.[91] Second, Savile introduced a source that Ramus never used in his history of mathematics: the fraudulent histories of Annius of Viterbo, supposedly written by the Babylonian chronicler Berosus.[92] From Annius, Savile cited some details about the landing of the ark and the subsequent settling of Noah and his progeny in Armenia.[93] Information drawn from Annius appears as well in Savile's section on Samothes and the Druids; though he seems not to have used these pseudo-historical fantasies anywhere else in his history.

Savile's remarks on the very beginnings of mathematics are quite sketchy; like Regiomontanus, he was far more interested (as he told his students) in those authors whose surviving works could be used to teach mathematics. Nowhere in the history itself did Savile even speculate on how or why mathematics was first discovered. There are, however, two passages in his preceding introductions to the mathematical arts that shed light on his thoughts on this problem.

The first passage occurs at the beginning of his discussion of arithmetic. According to Savile, this was the most natural of the sciences. It "was born when Man was born;" men come into the world with the ability to grasp its basic principles such as counting and simple addition, and as long as men exist, there will be arithmetic too.[94] If, Savile wrote, these inborn seeds of arithmetic are nourished by instruction, they will continue to develop until the student gains insight into the deepest properties of numbers. But the art of arithmetic itself was no human discovery. Unlike algebra, or *artificiosa arithmetica*, the ability to count, add, and subtract was a natural faculty of the human mind.

And yet, many ancient authorities had ventured accounts of the origin of arithmetic: Proclus attributed the foundations of arithmetic to the Phoenicians *propter*

---

[90] *Ibid.*

[91] *Ibid.* Savile was an implacable opponent of astrology throughout his life; see Goulding (1999, at notes 58–62).

[92] On Annius's forgeries, see Stephens (1984); Grafton (1991, pp. 76–103, especially p. 85); Ligota (1987).

[93] MS Savile 29, fol. 29r. The passage Savile was using is found at the very beginning of the third book of "Berosus," *De antiquitate Iani patris*; in Annius (1545), it is found at fols 22v–23r.

[94] MS Savile 29, fol. 8r: "Arithmeticam ... certum est ... una cum homine nato natam esse."

*mercaturas et commercia*, Josephus attributed it to Abraham, others to the Egyptians, and Diogenes Laërtius even believed that the Greeks had discovered it.[95] Savile did not attempt to resolve these divergent sources; that so many ancient authors had taken an interest in the question simply demonstrated the *importance* of arithmetic. Far more significant than the origins of the art, Savile concluded, were its two types of uses, for speculative philosophy and for practical needs. He then devoted the rest of his disquisition on arithmetic to an account of these applications.

Savile's description of the origin of arithmetic seems to imply that arithmetic is timeless and universal. His idea that this science was innate, – a natural art that the teacher merely had to nourish – has obvious and (for Savile) quite surprising Ramist overtones. It was, however, a point of view equally compatible with an orthodox Platonic account of the arts, in which all of the sciences were present somehow in the soul, waiting to be recalled. This was probably the sense in which Savile meant that arithmetic was innate. In his account of the origin of *astronomy*, which he delivered in a later lecture, Savile left no doubt of his Platonic leanings.

Savile's extended praise of astronomy and its origins is one of the most original and accomplished passages in the whole of his lectures. Again, Platonism and the contemplation of divine mysteries dominated his tribute to this science.[96] "According to Plato, the Homer of philosophers," humans were created "to contemplate the marvellous structure of the universe and the coordinated harmony of all things."[97] Savile found evidence for divine providence in every detail of the Ptolemaic universe. The earth occupies the central point of the universe; it was placed there in order to avoid the effects of parallax which would otherwise confuse our celestial observations. The earth is large enough for human habitation, but still is only a point in comparison with the heavens; this scale was chosen so that the greatest possible portion of the sky – one half – might be seen at any given time. In this way human beings were presented with an endless and constantly changing variety of heavenly phenomena.

But, Savile went on, humans were not meant just to admire the beauty of the stars. God put the stars in the heavens to stimulate our rational faculty, the part of our souls that distinguishes us from the animals. It was up to us to put reason to work, but God had provided ample signposts in the stars to guide us. In particular, there were countless celestial aids to the solution of the central problem of astronomy: the apparently random wanderings of the planets. To make these movements apparent to the primitive observer, God needed only to provide one or two fixed stars as stationary reference points. Instead, He filled the heavens with countless stars, grouped into the forms of men and animals, so that planetary movement would be even more obvious, and so that it could be conveniently recorded. God placed our nearest neighbour, the moon, so as to exhibit a regular pattern of phases: a simple phenomenon with a simple explanation, but finding the explanation was the first

---

[95] *Ibid.*, fol. 8v.

[96] What follows is all summarised from MS Savile 29, fols 18r–19v.

[97] Savile clearly means to recall *Timaeus* 47B.

step towards mathematical astronomy. The moon is also close enough to the earth to be susceptible to parallax. Since this presented the astronomer with great difficulties in formulating an accurate lunar theory, it might seem to be a poor decision on the Creator's part. But through the problem of parallax astronomers discovered the distance of the moon from the earth, and from there began to speculate on the size of the universe itself.

Savile adduced many more such examples, all of which led to a single, striking conclusion: astronomy was not invented for the stars, but the stars were made for astronomy. The Creator had provided the whole universe as a giant puzzle for humans to solve. Its very complexity was meant to arouse our sense of wonder – according to Aristotle, the root of all arts – and to stimulate in quite specific ways the development of the arts of the quadrivium.

Savile's great encomium of astronomy drew inspiration from Plato's *Timaeus*: the god gave humans eyes, said Plato, so that they could contemplate the heavens and thereby restore to themselves the use of reason.[98] From this conceit, we see how God set up the universe precisely so that modern astronomy might develop. Savile here combines something of Regiomontanus, who traced the discovery of astronomy to precise theoretical discoveries. But there was also a real attempt to make sense of the claim going back to Josephus, that the first human beings had, through long observation, built up astronomy from nothing. We recall that in his *Prooemium mathematicum*, Ramus had become less sanguine about this possibility than he had once been. Savile, on the other hand, did not find the claim incredible. The universe itself had been arranged to make such discovery possible, indeed inevitable.

## The Transmission of Mathematics

Like Regiomontanus, Savile was far more interested in the Greeks than in the legendary past. Still, he was careful to sketch out a chain of transmission from the earliest peoples through to the historical protagonists of Greek mathematics. Repeating Josephus's rationale for the long lives of the patriarchs, Savile wrote that Noah attained great skill in astronomy by virtue of the 950-year lifespan he had been allotted. After his arrival on the plains of Armenia, Noah's knowledge of mathematics spread to Persia, Babylon and the Chaldeans (the name given to the caste of Babylonian astronomers),[99] the last of whom surpassed in astronomy all the peoples who had preceded them. They were the first to fashion an astronomy that resembled the modern science, owing to assiduous observations made over many years – although not quite as many years as some had claimed for them. Savile was skeptical in particular of Diodorus's statement that Chaldean astronomers had 43,000 years of observations to draw upon, noting that none of the *scientific* writers who made use of Babylonian records (Aristotle, Hipparchus and Ptolemy,

[98] *Timaeus* 47A.

[99] MS Savile 29, fol. 29r.

in particular) said anything about such fantastically ancient records. Indeed, after Alexander's conquest of Babylon, the Greeks made off with all the Chaldeans' observational records; yet Callisthenes was able to send Aristotle only 1903 years of observations – no small achievement, but one that stretched back "only" to 100 years *after* the Flood, and not millennia before the Creation, as the Greeks (or the Babylonians) had pretended with "extraordinary impunity in their lies."[100]

Nevertheless, Savile was fascinated by Diodorus's account of the transmission of Jewish learning to the Babylonians. This seemed to him to be a crucial step in the migration of mathematics. The Chaldeans, having inherited their knowledge from Noah, formed one link in the chain that would eventually bring mathematics to the Greeks. But Savile also knew that the astronomers of Babylon had provided Hipparchus and Ptolemy with their most important observational data. In other words, the Chaldeans had informed the Greeks both indirectly and directly. Savile thus discerned a complex web of learning – all ultimately springing from the ancient patriarchs – which was artfully reunited in the genius of the Greek mathematicians.

Echoing Josephus, Savile related that Abraham formed the link between the Chaldeans and the next nation in the orderly migration of learning, the Egyptians. For Savile, Abraham was the most important figure in the early history of mathematics, acting not only as a teacher and transmitter between cultures (as he did for Josephus and so many who adopted Josephus's account, including Ramus), but also as a filter, so to speak, of the harmful accretions that had attached themselves to mathematics.

Abraham was a Chaldaean, wrote Savile, who came to realize that the stars were not gods, and was the first to assert the existence of a single god, creator of all the stars. Persecuted because of his unorthodox beliefs, Abraham moved first to Canaan (where he taught the Phoenicians, who subsequently developed mercantile arithmetic), and later to Egypt. The Egyptians at that time had no inkling of the sciences, and they eagerly absorbed Abraham's lessons in geometry and astronomy. Savile interrupted the Josephan narrative here to specify exactly what *kind* of mathematics he taught: it was, he said, "that pure, chaste and uncorrupt mathematics," cleansed of the kinds of "physical conjectures" and astrological predictions with which the Chaldeans had muddied it.[101] Abraham thus inherited the most advanced astronomy of the age – that of the Chaldeans, founded on the wisdom of Noah and subsequently improved by centuries of observation – but cleansed it of all the noxious, foreign elements that had been foisted upon it.[102]

It was this pure, advanced mathematics that flourished in Pharaonic Egypt. Later, the art continued to thrive, first under Alexander and the Hellenistic scientists of

[100] MS Savile 29, fol. 30r ("insignem mentiendi impunitatem"). On Callisthenes observations, see Grafton (1991, pp. 134–135).

[101] fol. 30v: "Mathematicam illam intelligo puram, castam, incorruptam, non physicis coniecturis, aut praedictionibus Chaldaicis contaminatam."

[102] The notion that the Chaldeans were responsible for spoiling ancient mathematics is also found in the writings of Pico della Mirandola (see Popper 2006, p. 91), Savile was influenced by Pico's anti-astrological arguments, and may have taken from him also this negative assessment of the Chaldean "additions" to astronomy.

Alexandria, then under the Romans, who were anxious that mathematics should never perish. In other words, from Abraham to Ptolemy to Theon of Alexandria, the art of mathematics had survived in Egypt for 2,500 years. It was finally destroyed by "Turkish cruelty" (i.e., the Islamic conquest of Egypt in the seventh century A.D.), when it perished along with all the other liberal arts. Then mathematics itself became homeless: "the holiest of all the arts was forced to go in search of new dwelling places."[103] Addressing the students directly (something he did quite rarely in his historical lectures), Savile concluded:

> But let us put aside this complaint. Let us rather concern ourselves with how we may bring Alexandria to Oxford – or, if this is too much to hope for, how the men of Oxford may understand the Alexandrian Ptolemy.[104]

Throughout this passage Savile emphasized the *Egyptianness* of Ptolemy and other Alexandrian mathematicians. The Greek identity of these scientists was not as important as the fact that they occupied a place in the pure tradition of Egyptian mathematics – a tradition which descended directly from Abraham and could thus claim to represent the most highly developed form of the patriarchal science, purged of the harmful elements ("physicality" and association with astrology) that it had picked up from the Chaldeans. It was a tradition looking for a home and, by more than coincidence, was precisely the *kind* of mathematics which Savile was convinced belonged at Oxford.[105]

Savile's exhortation to bring Alexandria to Oxford carried an extra level of significance, in that Ramus claimed in the *Prooemium mathematicum* that Alexandria had already found a new home, having been removed to *Germany*. The great patron of astronomy Wilhelm, Landgraf of Hesse-Kassel, had brought Alexandria to Kassel, a city distinguished for its many astronomical instruments, which were so advanced that "Ptolemy seems to have come from Egypt to Germany bringing armillaries and rules with him."[106] Ramus called Kassel "Alexandrian" because it had instruments comparable to those that Ptolemy might have used. For Savile, by contrast, the *theoretical* mathematics of Ptolemy was typically Alexandrian. If German astronomers (whom Savile admired just as much as Ramus did, but for quite different reasons)

---

[103] MS Savile 29, fol. 30v: "et omnium [artium] sanctissimam mathesim nova domicilia conquirere coëgit."

[104] *Ibid.*: "Illud potius agamus, quomodo possit Oxonium Alexandria transferri, aut si illud maius est, quam ut optari debeat, certe quomodo possit Alexandrinus Ptolemaeus ab Oxoniensibus intelligi." In the manuscript, Savile has underlined this passage heavily. He goes on afterwards to draw attention to his unwonted departure from the historical narrative, further emphasizing the significance of this passage. ("Sed prope oblitus eram me nondum ad haec tempora discendisse. Domum redeamus.")

[105] The mid-seventeenth-century Savilian professor of Geometry, John Wallis (who was familiar with Savile's manuscripts) used a similar argument to much the same ends in his own inaugural lecture. See Popper (2006, p. 104).

[106] Ramus (1569, p. 67): "Guilielmus Landgravius Hessiae videtur Cassellas Alexandriam transtulisse; sic Cassellis artifices organorum observandis syderibus necessariorum instruxit, sic quotidianis per instructa organa observationibus oblectatur, ut Ptolemaeus ex Aegypto in Germaniam cum armillis et regulis venisse videatur."

could be granted an observational Alexandria in their country, then Oxford would lay claim to the Alexandria that had sheltered *theoretical* mathematics in its purest and most advanced state – and Savile would bring this about by leading his students through the quintessential work of Alexandrian science, Ptolemy's *Almagest*.[107]

Savile clearly differed with Ramus on the question of what type of mathematics was best. He also took a different view of the shape and direction of mathematical history. As his treatment of the Chaldeans shows, Savile coupled his narrative of transmission to a *progressive* history, in which observational persistence or theoretical ingenuity could elevate the science to a much higher level than it had ever occupied before. For this reason, the latest phases of a science were at least as important and interesting as its origins, perhaps even more so. Although Savile acknowledged the astronomical accomplishments of Seth, Noah and other worthies, he also saw the Greeks, in particular, as doing something quite *different* from any of their predecessors. They may have drawn from the ever-branching streams of ancient wisdom, but they had transformed it into something new: above all, they introduced the notion of demonstration, built mathematics into a synthetic structure and made *new* discoveries, in each of these ways surpassing the work of the mythical ancients. In this progressive account of mathematics, we can clearly see the influence on Savile of Proclus's *Commentary on Euclid's Elements*. In Proclus's short history of mathematics, as the art passed from its barbarian discoverers to the Greek world, the art became progressively more abstract, and more distant from the real-world problems in which it had originated, such as restoring land after the flooding of the Nile. While Proclus acknowledged the barbarian origins of the sciences, he reserved their full development as abstract, demonstrated systems of knowledge for the Greeks. Savile agreed whole-heartedly; nowhere is his sympathy for this view more evident than in his biographies of ancient Greek mathematicians, beginning with the mystical sage who had received such unexpected attention in Ramus's *Prooemium*, Pythagoras.

In some ways, Savile's Pythagoras closely resembles the typical Renaissance portrait: an admirable, exotic, globe-trotting sage, who travelled among barbarians, studied with the Chaldeans, and brought back arithmetic, geometry and "sacred scriptures." from the Egyptians. He also journeyed far in mathematical contemplation, finally freeing the science from its material origins and exploring it purely through the activity of the mind. He made some fundamental mathematical advances: for instance, he constructed the first abacus and, of course, he discovered the theorem that now bore his name. Above all, wrote Savile, he deserved praise for his transformation of mathematics into a liberal art.[108]

Ramus, too, had considered this last achievement of Pythagoras (attested by Proclus) to be the most important. But Savile interpreted the meaning of Proclus's

[107] There is some irony in the fact that, several years later, Savile would work extensively with an astronomer who had enjoyed the Landgraf's patronage at Kassel, the "German Alexandria." His work with this astronomer, Paul Wittich, concerned entirely *theoretical* problems of planetary astronomy. See Goulding (1995).

[108] MS Savile 29, fols 32v–33r.

statement quite differently than Ramus did. While Ramus represented Pythagoras as a remarkably effective teacher, an author of textbooks, and founder of a school, Savile stressed his discovery of a chain of geometrical truths, arranged from the most general to the most particular. Ramus praised practice, Savile, theory. Moreover, Savile's depiction of Pythagoras formed part of a larger narrative which, in its structure, was quite incompatible with Ramus's history. For Savile, progress in mathematics and growing sophistication in demonstration went together. Demonstration was a good thing, and it was the specific contribution of *Greek* mathematics.[109]

In his life of Thales, Savile stated more than once that not only had he brought back geometrical facts from Egypt to Greece, but he had also been the first to provide demonstrations – *that* was the reason for the esteem in which he was held.[110] Savile's appreciation of this specifically Greek contribution to geometry was particularly apparent in his account of Hippocrates of Chios, whom he described as the "third father of geometry, after Thales and Pythagoras." His two predecessors had discovered the *principle* of demonstration.[111] Hippocrates developed the entire apparatus of demonstration in his early edition of an "Elements" of geometry, in which he first introduced proof by reduction to impossibility, an absolutely indispensable technique both for geometry and the other sciences. For this feat alone, he deserved to be considered the first *true* father of mathematics.[112] One can detect the influence of Ramus in Savile's elevation of *elementators* as the true founders of mathematics. But there is an important difference. It was complexity and sophistication that Savile admired in Hippocrates, not simplicity. Simplicity had been the principal concern for Ramus, who *regretted* that Hippocrates had ever invented the reduction to absurdity.[113]

Another example of the progressive drive towards complexity was Hippocrates's investigations into the quadrature of "lunes" – crescent-shaped figures formed from two circular arcs. By successfully squaring these shapes, Hippocrates was the first mathematician to discover anything of significance about a curvilinear shape.[114] Once again, Hippocrates had done something entirely different from his barbarian predecessors, extending the province of mathematical reasoning beyond the rectilinear. But above all he was to be honored for the introduction of demonstrative proofs – as Savile saw it, the *essence* of mathematics, not its decadence.

Ramus saw the history of mathematics as a story of gradual corruption from an original state of nature – this explained *why*, in short, modern mathematics was difficult and obscure when it (of all arts) should be natural and simple. Savile saw Ptolemy and other Greek mathematicians as occupying the end point of Egyptian

---

109 The role Savile assigned to the Greeks bears some resemblance to the historical model put forward by Pico della Mirandola in his work against divinatory astrology. See Popper (2006, p. 91).

110 MS Savile 29, fols 32r–v.

111 *Ibid.*, fol. 34v: "tertius post Thaletem et Pythagoram parens Hippocrates Chius ..."

112 *Ibid.*, fol. 35r.

113 See p. 66 above.

114 MS Savile 29, fol. 34v. For Savile's later researches into Hippocrates's quadratures, see Goulding (2005).

science, benefitting from and further improving upon the purified and progressive ancient tradition they had received from their Egyptian forebears. For Ramus they were protagonists in a peculiarly Greek story of degradation, in which natural reasoning had been abandoned in favor of the artificial and hypothetical.

Inevitably, then, Ramus's program for mathematical education looked very different from Savile's. Savile wished to recreate the mathematics of Greco-Egyptian Alexandria in Oxford; according to Ramus, mathematics would only be restored to its ancient purity by cutting out the Greek authors altogether (or, at least, by emulating them only insofar as they themselves pursued practical mathematics). Instead, Ramus thought, the natural, unfeigned science would be found by scrutinising the actions of merchants and artisans, who unknowingly practiced the same art as the uncorrupted patriarchs, drawing it, as *they* did, from the natural resources of the human mind. It was this natural, practical, and unaffected science that Ramus wished to have taught at the University of Paris.

## Savile's Platonism

Savile's emphasis on the theoretical progress of mathematics over time went hand-in-hand with a sometimes exaggerated Platonism which he affected throughout his lectures. Savile tried constantly to present himself as an other-worldly Platonist, at one point refusing even to consider applications of the sciences: "due to a sort of disgust for external things (he wrote) my mind has always shrunk from consideration of [such subjects]."[115] Rather, it was the abstract, theoretical side of mathematics that drove Savile ever on in his studies, at times transporting him into powerful, mystical ecstasies. Savile told his students that it was in the second year of his studies[116] that, immersed in the works of Euclid, Ptolemy and Archimedes, he experienced a sudden insight into the beauty and harmony of theoretical mathematics. "My soul was flooded with intense pleasure, and I sought nothing else in the way of utility or diversion."[117] Or, as he put it elsewhere in his lectures, "I am passionately inflamed with love for geometry. Even here, even now, I can barely restrain my tumultuous emotions."[118] Such emotive outbursts would have been remarkable in any context; it hardly needs to be said how unusual they must have been in the pedestrian and conventional genre of the ordinary lectures.

[115] MS Savile 29, fol. 11r: "quanquam ab huiusmodi commemoratione nescio quo rerum externarum fastidio semper animus meus abhorruerit."

[116] Presumably his MA studies, when he would have been reading geometry.

[117] MS Savile 29, fol. 5r: "... ut altero iam anno quo animum ad discendum inieci, et in Euclidem, Ptolemaeum, Archimedem praecipue incubui, prope incredibili perfusus animi voluptate, nihil extra quaesiverim vel ad utilitatem vel ad oblectationem."

[118] MS Savile 29, fol. 10v: "Geometriae tamen amore intemperanter ardeo, nec immoderatos hoc loco eos impetus facile cohibeo."

But this was more than a personal quirk; Savile's Platonic program was one strategy for recommending mathematics to his students and to the University as an institution. Savile provided many examples – both in the introductory lecture and, especially in the later history of mathematics – of ancient figures, heroes of the humanist movement, who had devoted themselves to the study of the sciences. But his central thesis in the early lectures was that mathematics was an essential part of philosophy – Platonist philosophy in particular. Complementing this was an insistence that mathematics had nothing whatsoever to do with merchants and common craftsmen; as modern scholars have suggested, one of the most significant obstacles for the sciences at Oxford may have been the perception that they were not suitable subjects for a gentleman.[119] In his discussion of astronomy, for instance, Savile touched very briefly on the subject of navigation, but abandoned it quickly with the remark that "I know most of you despise it."[120]

Mathematics, he insisted, was a pure philosophical art: a "path to theology," releasing its practitioner from the fetters of Plato's cave. His audience knew how much effort he had expended in mastering these arts; did they think he would have spent so many sleepless nights poring over Archimedes and Ptolemy if he had only wanted to be able to measure the height of a tower or construct an astrolabe? These practical tasks did involve mathematics, he admitted; but the theoretical mathematician descended from his contemplation to instruct and guide the hand of the unlettered craftsman; he did not dabble in such things himself.[121]

The class-consciousness of the new breed of Oxford student has been well-documented. It is a little puzzling, though, that the kind of skills Savile dismissed in this passage – making an instrument or measuring a tower – were precisely the accomplishments which gentlemen did in fact find attractive in the sciences. If anything, the *theoretical* pursuit of the sciences was considered unbecomingly bookish. It might seem, then, that Savile was deliberately inverting the expectations of part of his audience. But we should keep in mind that Savile was addressing not young undergraduates, but MA students; astronomy was reserved for students who had already received the bachelor's degree. These students had perhaps more intellectual ambition than the young noblemen who sought only some polish from a year or two of university, and often left Oxford without a degree. In other words, he may have been identifying himself and his audience as the *real* scholars at Oxford, whose *bona fides* would be established by their mastery of mathematics purely for its own sake.

Moreover, Oxford was generally predisposed to reject anything that smacked of Ramism – and it may even be that the publication of Ramus's *Prooemium*

[119] Feingold (1984, pp. 190–192); Taylor (1954, pp. 4–5). Pumfrey (2004) makes the point that the worlds of the universities and the practitioners were more widely separated in England than in anywhere else in Europe: the kind of *bricolage* common on the Continent, where mathematicians moved back and forth between these sites of knowledge and refashioned their identities as they went, was almost unknown.

[120] MS Savile 29, fol. 22r–v: "... ne semper haeream in navigationibus, quas scio plerosque vestrum contemnere ..."

[121] MS Savile 29, fol. 5r–v.

*mathematicum* had stirred up a distaste for mathematics at Oxford: on the grounds that if Ramus liked it, it could hardly be worth studying. It is notable that, although Savile attacked Ramus many times in his lectures, he never once stopped to explain to his students who Ramus was; nor did he ever refer to the *Prooemium mathematicum* directly, even while admonishing Ramus for his philosophy of mathematics. He seemed to assume that his students knew precisely who Ramus was, and were inclined to dislike him. Indeed, Ramist disputes were public and vocal; it was hardly possible to pass through one of the English universities without being aware of Ramus.[122]

Thus Savile's exaggerated idealistic Platonism and his almost exclusive support for theoretical mathematics may have been encouraged by his desire to present himself as a scholar as different as possible from Ramus. From a lofty Platonic vantage point, he castigated Ramus for his practical concerns, exclaiming: "Immortal gods! That mathematics – until now immune from all thoughts of worldly advancement – should be reduced to a mere mechanical skill, as though thrust into some lowly mill!"[123] And he left no doubt of the reasons for Ramus's inadequacies in his most extended attack on Ramus in his lectures:

> I wish, Ramus, that you had followed Plato, Ptolemy, Proclus, and other philosophers more learned than yourself and declared contemplation of the eternal realities to be the purpose of the most liberal of arts; instead you made the purpose of these arts mechanical and illiberal.[124]

Any self-respecting craftsman, Savile continued, would be ashamed to be seen trying to build a wall without his level and plumbline. How then could Ramus, a master of mathematics, go about his work without the help of Euclid or Pappus? Ramus's disregard for philosophy only demeaned himself:

> If only you had extolled the ascent to separated substances with your unique rhetorical gifts and had vigorously demonstrated your eloquence to us – now instead we despise it, immersed as it is in such baseness.[125]

It is ironic that Savile compared Ramus unfavorably with *practical* mathematicians, such as those who used instruments to build walls. These were precisely the kinds

---

[122] See Feingold (2001). There is clear evidence that, a few years after Savile's lectures, Ramism was a matter of public discussion in Oxford. In 1583, shortly after Savile returned to England from his European tour, a letter written home by an Oxford undergraduate recorded that Savile was widely tipped to be the university's champion in a refutation of the Ramist philosophy as advanced from Cambridge by William Temple (MS Oxford, Bodleian Library, Rawlinson D.985, fol. 52v; edited in (Jeffery, 1909), p. 57).

[123] MS Savile 29, fols 5v–6r: "Dii immortales, mathesin quae hucusque ab omni vitae commodo sacrosancta fuit ad mechanicae tractationem tanquam vilissimum aliquod pistrinum detrudi!"

[124] MS Savile 29, fol. 5v: "Hunc tu finem artium liberalissimarum, Rame, cum Platone, Ptolemaeo, Proclo, et reliquis doctioribus philosophis constituisses velim, non illum mechanicum et illiberalem."

[125] MS Savile 29, fol. 5v: " Istam si tu pro tua singulari in dicendo facultate ad substantias separatas *anabasin* exornasses, et tuam nobis vehementer probasses eloquentiam, quam nunc quidem in tanta rerum foeditate contemnimus."

of practitioners to whom Ramus looked for the restitution of mathematics. In fact, however, by rejecting the *theoretical* foundations of the applications of mathematics, Ramus could not even aspire to the proficiency of one of the craftsmen that he so admired.

## Savile's Platonic Biography

To some degree, Savile's professed Platonism was conventional. It was a commonplace of Continental introductions to mathematics to adopt this lofty view of the sciences, as we have seen in the orations of Cardano and Melanchthon. But the lengths to which Savile took his Platonism were quite extraordinary. At one point in his lectures, he gave his students an account of his own education. Here Savile said nothing about his wide reading of contemporary Continental practitioners (enterprising and impressive as this had been). Rather, he fashioned an intellectual autobiography that entirely conformed to a Platonic ideal. He claimed that his interest in mathematics had been stimulated by a "certain man" or "guiding spirit," who set him to working his way through Euclid's *Elements*. Savile described his book by book progress:

> Who would not be delighted by so noble, so pleasing, so agreeable a variety of the most pleasant of things? I was greatly moved by the certain and established passages from first principles to intermediate results, and from them to advanced theorems, set out in a straightforward order of progression. It was pleasant to embark on triangles. Debates on quadrilaterals pleased me greatly. I delighted first to describe circles, and then to unite them with the aforementioned figures. Then, before I had grown tired of circles, the sweet harmony of proportion seized me, in which I should always wish to linger – but I did not wish to die having mastered so little of geometry. And so, not irrationally, I applied myself to irrationals. Why go on? I retired exhausted. I had scarcely made my first acquaintance with stereometry when I bade farewell to geometry.[126]

Abandoning the *Elements* after the rigors of its tenth book, Savile set about reading the *Almagest*, yet found himself unable to make any headway whatever. Realizing that he did not have sufficient geometrical background, he returned to the *Elements*, and worked through the last three books on solid geometry: the construction of cubes, pyramids, dodecahedra and so on. Now properly equipped, he was able to master planetary astronomy.

---

[126] MS Savile 29, fol. 11v: "Quem enim non delectaret tam illustris, tam grata, tam iucunda rerum suavissimarum varietas? Me certe vehementer affecit certas esse et statas vices a primis a media, a mediis ad ultima directo quodam ordine proficiscendi. Iuvit in triangulis pedem ponere. De quadrilateris disceptationes mire placebant. Delectabat iam versasse circulos, iam ad praedicta comparasse. Nondum de circulis defessum excepit suaviss. proportionum concentus, in quibus morari semper vellem nisi mori parum geometer noluissem. Itaque irrationalia non sine ratione attigi. Quid multa? Fatigatus discessi. Stereometriam demum vix a primo limine salutans, geometriae vale dixi."

Despite the claim that it was a personal history, Savile's account here was in fact highly contrived, and probably bears no relationship at all to his actual education.[127] The pretext for abandoning the *Elements* should, in particular, arouse our suspicions. The reader of the *Almagest* has no need to understand the construction of polyhedra; having completed the books on plane geometry and arithmetic, Savile was as well prepared as he could be to understand ancient astronomy.[128] Why invoke such an unlikely stumbling block? The motive behind Savile's strange assertion may perhaps be found in Plato's *Republic*. In book VII, Plato has Socrates and Glaucon discuss how the education of the philosopher king must progress from arithmetic to plane geometry and thence to astronomy, the study of "solid bodies in motion." Scarcely had they begun to consider this subject, however, when Socrates called Glaucon to a halt: "... you must go back a bit, as we made a wrong choice of subject to put next to geometry.... We proceeded straight from plane geometry to solid bodies in motion without considering solid bodies first on their own. The right thing is to proceed from second dimension to third, which brings us, I suppose, to cubes and other three-dimensional figures."[129] Only then do they return to astronomy.

Savile evidently shaped his account of his own education in order to conform to the ideal education of the philosopher – even to the extent of preserving a "premature" misstep into astronomy. By modelling his intellectual biography on the *Republic*, Savile was invoking the support of the greatest contemplative philosopher of antiquity; his enthusiastic embrace of Plato came in stark contrast to Ramus's rejection of Plato in the *Prooemium mathematicum*. And whether or not his students recognized the allusion, they could still see that Savile had not been motivated in his study by any particular application of mathematics; they, too, were being asked to look beyond Ramist utility and to pursue mathematics simply for its own sake – or for the superior kind of joy and pleasure that Savile himself experienced in his journey through the *Elements*. And, finally (and most cleverly), Savile's account of his education was a deliciously ironic reworking of Ramus's own intellectual autobiography. Ramus, too, had been deterred by the formidable tenth book of the *Elements*.[130] But the French philosopher did not pursue the ideal education of a

[127] Savile's personal copy of Euclid is in the Bodleian Library with shelfmark Savile W.9(1) (Euclid, *Stoicheia*, ed. Simon Grynaeus, Basel, 1533). The volume is heavily annotated, but offers no support for Savile's account. Note that Savile makes no mention of the arithmetical books of the *Elements*; his own copy of Euclid, however, suggests that he read them continuously with the geometrical ones.

[128] At fol. 9r Savile told his students that the geometrical preparation needed for studying the *Almagest* was the first *six* books of the *Elements*, scarcely anything from the tenth book on irrationals, as well as a theorem or two from the eleventh and thirteenth books (both of which books concern stereometry). From the eleventh book, he no doubt meant the propositions on intersecting planes; it is difficult to see what he might have required from the thirteenth. In any case, Savile was hardly demanding from his students the curriculum that he said he had found essential in his own education. ("In geometria sex primos elementorum libros, et nonnulla undecimi, decimique tertii theoremata, vix decem.")

[129] *Republic* 528A–B.

[130] See passage cited at n. 49 of second chapter.

philosopher king, nor later return to finish his journey through Euclid, as Savile did. On the contrary, Ramus entirely disparaged ancient mathematics, proposing instead his own simplified and utilitarian mathematics to take its place.

## Ramus, Savile and English Science

Savile's animus towards Ramus was not entirely founded on philosophical differences; national pride was also at stake. Throughout the *Prooemium mathematicum*, Ramus had called for a mathematical revolution at the University of Paris – a development that would redound to the glory of the French crown and nation. As he stressed in his survey of European mathematics and mathematicians, France lagged far behind the rest of Europe, Germany and Italy in particular. The only nation the French could be confident of surpassing was the English, who found themselves in an even worse state than the French. In his *Prooemium*, Ramus addressed the English queen on the future of mathematics in her nation:

> But Elizabeth, Queen of England, do not allow your England to be a pupil of France any longer, but summon the French in their turn over to England... Inquiring into the two most erudite universities of your realm, I have learnt that professors of Greek, Hebrew, medicine, civil law, and theology are honoured with royal stipends... but no royal reward has been established for professors of mathematics... And so for you, Your Majesty, I desire Regius Professors of mathematics in both Cambridge and Oxford, to adorn your memory with eternal praise for your magnificent generosity.[131]

Savile had Ramus's words in mind when he lamented a "recently published criticism" of mathematics at the universities, that was so damaging to their reputation.[132] Savile may have found Ramus's unsolicited advice to the Queen galling, but he had to admit that the French philosopher had identified a deficiency in the English academies. He himself had met with influential men (Savile told his students) who wanted to know of two mathematicians, one at each university, worthy of their support: they intended, claimed Savile, to award them regius professorships. Savile was unable to suggest anyone, and sent these would-be Maecenases away

131 Ramus, *Scholae mathematicae*, p. 14: "At Elizabetha Anglorum regina, Angliam tuam Galliae discipulam diutius fieri ne sinito, sed Gallos vicissim in Angliam provocato... In duabus eruditissimis regni tui academiis sciscitando didici regiis stipendiis honorari professores linguae graecae, hebraicae, medicinae, juris civilis, theologiae... At mathematicis artibus praemium regale nullum est constitutum... Itaque opto regios reginae Elizabethae in academia et Cantabrigiensi et Oxoniensi mathematicos professores, qui sempiterna praeclarissimi beneficii laude memoriam tuam exornent." One of Ramus's correspondents about the state of mathematics in England was John Dee. In a letter to Dee, written in 1565, that was published in Ramus (1599, pp. 174–175), Ramus asked "who in your universities teaches mathematics, and with what authority?" ("quinam in vestris gymnasiis, quaque authoritate mathematicas artes profiteantur.")

132 MS Savile 29, fols 2v–3r: "quam quidem ingratam et adversam dignitati nostrae famam, scripta non iam pridem severa monitione auctam et amplificatam vidimus."

disappointed.[133] Savile's odd story completes and subverts Ramus's account of his researches into the state of English mathematics. While Ramus suggested to the Queen that she should found regius professorships in mathematics, as if no one had thought of this solution before, Savile claimed that there was *already* interest in England in just such a foundation, but there were no mathematicians to fill the positions.

Savile bristled at Ramus's audacity. His ordinary lectures served as a kind of answer to Ramus's condescending concern for England. Oxford and Cambridge might not have regius professors in the sciences, but Savile's boldly ambitious ordinary lectures would be at least the equal of any royal lecture – surpassing, in particular, those of the regius professor in Paris, Peter Ramus. And instead of decrying the contributions of classical mathematics and calling boldly for the dismantling of the Greek tradition, Savile surveyed the history of his discipline with a much more conventional Renaissance spirit. If mathematicians would just read their ancient texts, he argued, mathematics itself would be restored, and Alexandria rebuilt among Oxford's quadrangles and spires.

## Conclusion

Savile's lectures are a curious mixture of philosophical conservatism and an ambitious reforming spirit. The uneasy balance reflected his ambivalence about Ramus. With Ramus he shared a conviction that something needed to be done about the state of mathematics in the academy. And like Ramus, Savile attempted to formulate a solution to the predicament by reviewing its historical genesis, expressing the nature and utility of mathematics through his account of its development – an account which, in many of its details, he took directly from Ramus's history. Yet Savile intended his own narrative to support a philosophy of mathematics entirely opposed to that of the French logician, who irritated him as much for his presumption in taking on mathematics in England and Paris single-handedly, as he did for his utilitarian approach to the sciences. As I have suggested, Ramus's work was a provocative goad to Savile's own intellectual development; his resolute insistence on practicality drove Savile to adopt an uncompromising Platonic account of mathematics and its ends.

Savile wrote his *Proemium* as a rebuttal to Ramus's, even as he made extensive use of the material Ramus had compiled. The irony of this situation becomes particularly apparent when one considers the historical narratives that the two men developed. Savile used essentially the same historical data as Ramus, and, to a large extent, he took it directly from Ramus himself; yet he came to opposite conclusions, both about the historical protagonists involved in the narrative and about the *implications* of that narrative for the nature of mathematics. A point that I have made

[133] MS Savile 29, fol. 7v.

throughout this book has been that historical narratives were quite malleable: just as Ramus could bend the biographies of Pythagoras and Plato to support a Ramist model of mathematics, so too could Savile take the same evidence to construct an anti-Ramist, Platonic account.

It is interesting to contrast the high Platonism of the youthful Savile with the sober balance he struck between theoretical and practical mathematics in the statutes for his two professorships half a century later. When he endowed the Savilian chairs in 1619, three years before his death, Savile drew up an eminently sensible set of statutes, governing the appointment of the lecturers and their duties with respect to both teaching and research (the latter a novelty in the English academy).[134] Their research, Savile stipulated, should above all comprise the study of classical Greek mathematics, although they should not neglect modern advances. The professors themselves should not remain content with explicating and illustrating the work of others, but should try to develop and enlarge their disciplines – in the case of the Professor of Astronomy, by nightly celestial observations. To this end, they must deposit their findings in the library for the use of future Professors, just as Savile had done with his own notebooks. As a research programme, Savile's specifications carefully balanced scholarly, humanistic study with mathematical specialization – specialization that required, though only to a modest degree, familiarity with mathematical practice.

Among his Professors' teaching responsibilities, Savile again makes most provision for the exposition of the theoretical part of the disciplines – yet he does not, by any means, neglect practical mathematics. The Professor of Astronomy's primary teaching duty was to expound the entire *Almagest* of Ptolemy, supplementing it with Copernicus, as well as with other works ancient and modern – in other words, to lecture in much the same way that Savile himself had done half a century before. To assist his successors in emulating him, Savile included his lecture notes in the Professors' mathematical library. But Savile also required the astronomy professor to teach his students the whole of optics, dialling, geography and "those parts of navigation which are founded on mathematics," so long as the university would release him from other duties.

The Professor of Geometry was also required to teach subjects beyond the conventional, pedagogical bounds of his discipline. According to Savile's directions, he must lecture on all thirteen books of Euclid's *Elements*, the entirety of Archimedes's works, and the *Conics* of Apollonius. Like his astronomical colleague, the geometry professor also had some practical duties. In addition to his regular teaching of the mathematical classics, the Professor was to teach his students geodaesia (that is, "practical geometry" or surveying), music and mechanics. Perhaps reflecting Savile's low estimation of the abilities of the average Oxford undergraduate of the time, Savile directed the Professor of Geometry to teach, at least once a week, simple arithmetic in his rooms, in English if necessary. If this were not enough, he

[134] The statutes are reprinted in Gibson (1931, pp. 528–540). The following paragraphs are summarized from pp. 528–529 and Sect. 5 on p. 531. On the rarity of the notion of "research" as part of a professor's duty, see Curtis (1959, p. 227).

stipulated finally that the Professor should "at suitable times, convenient to him, demonstrate the practice of geometry to those of his students who wish to attend, in places around town or in the nearby countryside."

It would be wrong to overstate Savile's enthusiasm for applied mathematics as he drew up his statutes. In setting out his criteria for the selection of the professors, he said not a word about their practical abilities. Rather, appointees were to be highly skilled in mathematics and thoroughly educated in philosophy, through reading the works of Plato and Aristotle (and *not* the commentaries of the scholastics). They should also know some Greek.

Towards the end of his life, in fact, Savile became infamous for his curmudgeonly distaste for practical mathematics. John Aubrey told the well-known story of Savile interviewing the unfortunate Edmund Gunther for the position of the first Savilian Chair of Geometry.

> So [Gunther] came and brought with him his Sector and Quadrant, and fell to resolving of Triangles and doeing a great many fine things. Said the grave Knight, Doe you call this reading of Geometrie? This is shewing of tricks, man! and so dismisst him with scorn, and sent for Henry Briggs, from Cambridge.[135]

Savile was still enough of a theoretician to distrust practical achievements pursued (as he saw them) for their own sakes. Nevertheless, the Savilian statutes show that, at least in more temperate moments, he acknowledged a place for the applications of mathematics – if only as useful, concrete tools for instilling in youth the underlying, theoretical sciences.[136] Savile's mature balance of the theoretical and the practical, of research and teaching, of astronomy and geometry – this, more than either his or Ramus's *Prooemium mathematicum*, suggested how the mathematics of Alexandria would be built anew in the modern age.

---

[135] Aubrey (1958, p. 268).

[136] See Goulding (2002), where I suggest that Savile's partial change of heart may be connected with the criticisms of the universities put forward by his chosen successor, Henry Briggs, and the perceived threat in the foundation of Gresham College.

# Chapter 5
# The Puzzling Lives of Euclid

## The Veiled Figure

According to a story told by the Roman antiquarian Aulus Gellius, Euclid of Megara exhibited a singular passion for philosophy. At the outset of the Peloponnesian War, Athens imposed sanctions against the nearby city of Megara and banned its citizens from entering Athens. Euclid, a student of Socrates, was distraught to be kept away from his master's debates, and came up with a clever subterfuge: each night, under the cover of darkness, he crept into town dressed in women's clothing and joined his fellow philosophers to hear their teacher's discourses, leaving the city again before dawn.[1]

Euclid went on to found the Megaric school of philosophy, renowned for their delight in paradoxes. One of the best-known puzzles associated with this school was that of the "veiled figure." If you are introduced to a veiled figure of your father, the paradox goes, do you know your father? Should you answer "yes," you are refuted on the grounds that the figure was veiled, and hence unknown when introduced to you. The answer "no" provokes the response that you certainly do know your own father.[2]

It is most appropriate that Euclid (or his students) should have devised such a paradox. Socrates, too, surely knew his student, yet did not know him when he arrived in Athens in disguise. This is not mere quibbling, but a real epistemological puzzle. The veiled figure and Euclid are the same person, so that Socrates could both "know" and "not know" a numerically identical figure. The paradox becomes even sharper if we ask whether Socrates knew that the veiled figure had blue eyes (assuming that is the color of Euclid's eyes). In the same breath, Socrates could assert both that he does not know whether the veiled figure has blue eyes, and that he does know that Euclid has. It is quite fitting that Plato made Euclid the narrator of his *Theaetetus*, a dialogue whose argument revolves around paradoxes of knowledge and identity.[3]

---

[1] Aulus Gellius, *Attic nights* VII.10.

[2] On this paradox and its association with Euclid of Megara, see Wheeler (1983).

[3] See Sorensen (2003, pp. 71–74), on Euclid, the identity paradox and the *Theaetetus*.

R. Goulding, *Defending Hypatia*, Archimedes 25,
DOI 10.1007/978-90-481-3542-4_5, 

And how ironic, from our point of view, that Euclid of Megara should be so engaged with these problems and paradoxes of identity. For, beneath the disguise of this wily, argumentative philosopher, Renaissance historians believed they knew just who it was who had crept into Athens by night. Euclid of Megara was, they supposed, none other than the author of the *Elements* of geometry. It was, admittedly, a puzzle that neither Plato nor Euclid's later biographer Diogenes Laërtius ever mentioned that Euclid of Megara had been a mathematician, despite the fact that they wrote of the man whose name was *synonymous* with geometry! To omit any mention of Euclid's single most important accomplishment was, in itself, a peculiar sort of concealment. It took some effort and more than a little faith to look beneath the veil and find there the author of the *Elements*.

The Renaissance unmasking of Euclid was, it turns out, entirely wrong. Euclid of Megara did *not* write the *Elements* of geometry, which was the work of an author who shared his name but lived nearly a century later. In an ironic inversion of Euclid's veiled man paradox, historians "knew" a disguised figure whom they did not, in fact, know at all.

## Towards a Biography of Euclid

In 1505, the Venetian humanist Bartolomeo Zamberti published his monumental Latin translation of the works of Euclid, containing not only the *Elements*, but also the *Phaenomena*, *Optics*, *Catoptrics* and *Data*.[4] In his preface, Zamberti acknowledged that his readers would likely want to know something about the author of these works, styled "Euclid of Megara, the Platonic Philosopher" on the title page. Yet, Zamberti confessed, he had been able to find little information on Euclid in any ancient text, and no comprehensive narrative of his life whatsoever. The best Zamberti could do was to reproduce without commentary the material he had discovered. The excerpts he published, gathered from various ancient biographies and philosophical and literary texts, seemed to describe a philosopher, a student of Socrates, who had turned the fragmented mass of early Greek mathematical knowledge into a coherent system and had played a crucial part in Plato's mathematical education and philosophical development. But, Zamberti cautioned, readers would need to work hard to make sense of the story, for the Greek and Latin authorities he quoted often conflicted with one another, and "it is not my place to make rash judgments about the writings of such great men."[5]

Zamberti's reticence reflects more than just the conventional humanist *topoi* of modesty and deference to antiquity. In fact, he had good reason to be puzzled by the texts he had gathered. For, despite his confident assertion on the title page that Euclid of Megara, the "admirable Socratic philosopher" was the author of the *Elements*, and thus "gatekeeper of the mathematical arts," Zamberti had, in fact,

---

[4] Euclid (1505). On Zamberti (c. 1474- after 1539), see Rose (1976, especially pp. 301–302).

[5] *Ibid.*, fol. 6v: "Nam nostrum non est de tantorum virorum scriptis ausu temerario iudicare."

gathered biographical data on two entirely different people, separated by nearly a 100 years. Euclid of Megara was, indeed, a philosopher associated with Plato and Socrates. In the *Phaedo*, Plato himself recorded that Euclid of Megara was present at Socrates's death; Plato also cast him as the narrator of his *Theaetetus*. By contrast, very little is known about Euclid the *mathematician*, except that he lived at least a century after Plato. On this basis alone, we can be quite sure that he was not Euclid of Megara.

Zamberti was not the first to conflate the two Euclids. There seems to have been some confusion about their identity even in antiquity. The first-century Roman rhetorician Valerius Maximus, in his collection of *Memorable Deeds and Sayings*, devoted a chapter to anecdotes illustrating the value of consulting with experts in which Euclid made a significant appearance. Valerius told how the priests at Delos, anxious to dispel a plague, consulted the oracle and were told to double the size of their cubical altar. Seeking advice on how to do this, they approached Plato, who

> told the keepers of the sacred altar to consult Euclid the geometer, yielding to his knowledge – indeed to his profession.[6]

Valerius was careful to specify that the Euclid whom Plato knew was a geometer by "profession," not just by virtue of his knowledge of mathematics. His story concerned a Euclid who was known *primarily* as a mathematician. Later readers assumed that Valerius meant the Euclid who wrote the *Elements*, though there is nothing in Valerius's story that identifies him explicitly so.

Valerius Maximus was not a mathematician nor even a historian, but a compiler of morally-improving anecdotes. His confusion of Euclid the friend of Plato with Euclid the mathematician was recognized as an error in antiquity. One of his ancient commentators, Mitalerius, emended the name "Euclid" to "Eudoxus," a mathematician who was indeed a contemporary and friend of Plato. The commentator was most likely aware of the version of the story preserved by Plutarch (a much more serious authority on philosophical and historical matters), according to which Plato advised the questioners to consult with either Eudoxus of Cnidus or Helicon of Cyzicus, two mathematicians of his Academy. This was surely the original form of the anecdote.[7] How Valerius managed to turn "Eudoxus" into "Euclid" is another question. The name "Euclid" had become so identified with the author of the *Elements* that Valerius must have assumed that the Euclid of Plato's dialogues was the mathematician to whom the philosopher turned for geometrical aid – rather than to those more obscure mathematicians, genuine contemporaries of Plato, who had originally been the subjects of the anecdote.

---

[6] Valerius Maximus, *Facta et dicta memorabilia*, VIII.xii ext. 1: "Platonis quoque eruditissimum pectus haec cogitatio attigit, qui conductores sacrae arae de modo et forma eius secum sermonem conferre conatos ad Eucliden geometren ire iussit scientiae eius cedens, immo professioni."

[7] Plutarch, *De genio Socratico*, 579B–D. See Euclid (1926, vol. 1, p. 3); Heiberg (1882, p. 23); and Knorr (1986, p. 2).

Valerius is the only ancient writer to confuse the two Euclids. Yet, judging by the common conflation of Euclid of Megara and Euclid the mathematician in Byzantine sources, there must have been a more extensive tradition, now lost. The fourteenth-century scholar Theodorus Metochites made the identification quite explicit, as he referred to

> Euclid of Megara, the Socratic philosopher, a contemporary of Plato, who brilliantly gathered together most of the geometrical results of that time.[8]

Several medieval Greek manuscripts which made their way into the West also attributed the *Elements* to Euclid of Megara. Influenced by their testimony, or by that of Valerius Maximus or, as seems most likely, by the notice in Theodorus Metochites – the Venetian printer Erhard Ratdolt introduced the notion to print. In his 1482 *editio princeps* of Campanus's Latin version of the *Elements*, Ratdolt concluded his prefatory letter to Doge Giovanni Mocenigo by praising the works of "Euclid of Megara... who perfectly gathered together the whole science of geometry in fifteen books."[9] After Zamberti published his collection of biographical sources on "Euclid," in 1505, almost all later editors of the text concurred that Euclid of Megara had written the *Elements*. The first English edition, published in 1570 by Henry Billingsley and John Dee, celebrated on its title page "the most auncient Philosopher Euclide of Megara."[10]

Eventually, however, Renaissance scholars began to discover contradictions in the traditional account and, as a result, ventured the claim that the *Elements* had been written by another Euclid, one who lived sometime after Plato. These scholars (chief among them, Peter Ramus) reached this conclusion after much laborious collation of prosopographical and chronological data, derived from an extraordinarily wide array of ancient texts. Part of my task in this chapter is simply to retrace their detective work. But the discovery of Euclid's true identity was not merely a story of historiographical error and humanist ingenuity. As I have argued throughout this book, Renaissance historians of mathematics often wrote to serve ideological ends: to clarify the nature of mathematics, to establish its purpose and role in the academy, to promote its value to the broader humanistic culture. The delicate surgery needed to separate Euclid the philosopher from Euclid the mathematician – impressive as it was in its own right – was performed against the backdrop of rather more robust Renaissance debates over the relationship between mathematics and philosophy (particularly of the Platonic school) and the proper domain of each.

---

[8] Cited in Heiberg (1882, p. 24); and Euclid (1926, vol. 1, p. 3), where Theodorus is said to be the first person (besides Valerius) to confuse the two Euclids.

[9] Euclid, Elements (Venice: Erhard Ratdolt, 1482, fol. 1v: "Euclides igitur Megarensis, serenissime princeps, qui xv libris omnem geometrie rationem consummatissime complexus est." The similarity of Ratdolt's phrase "consummatissime complexus est" to Theodorus's description of Euclid's activity is quite striking.

[10] See Heiberg (1882, p. 24), for a list of editions attributing the Elements to Euclid of Megara; and Thomas-Stanford (1926) for early editions of Euclid in general.

## Zamberti's Collection of Sources

As we have seen, Zamberti was not the first to attribute the *Elements* to Euclid of Megara, the companion of Plato. He did nothing out of the ordinary when he sang the praises of the "Platonic philosopher" on his title page. Yet his preface to the *Elements* marks a significant juncture in the history of the biographical error. For the first time, a comprehensive range of texts about "Euclid" had been gathered into one place. While earlier writers may have attributed the *Elements* to Euclid of Megara, none had attempted to consider the evidence for his identity, or indeed shown any interest at all in the author's biography. Zamberti's effort testifies to a new interest in the history of mathematics at the turn of the sixteenth century. Further, in his awareness that there was something peculiar about his sources, Zamberti's work marks the beginning, at least, of the sixteenth-century reassessment of the identity of Euclid – a reassessment that would eventually lead to a critical approach to the text of the *Elements* itself. Finally (and somewhat ironically), by identifying and citing in full almost all the texts related to one or other of the Euclids, Zamberti made it possible for later scholars to realize that the friend of Plato who also systematized geometry was nothing more than a chimera. Why Zamberti did not recognize this himself is something I will consider presently.

Zamberti's collection of biographical extracts, appearing as an appendix to his preface, presents almost all of the significant texts on "Euclid," whether the philosopher or the mathematician. These can be summarized as follows:

| | Text | Which Euclid *really* |
|---|---|---|
| 1 | *Suda* | Megara |
| 2a | Diogenes Laërtius (on Euclid) | Megara |
| 2b | Diogenes Laërtius (on Socrates) | Megara |
| 2c | Diogenes Laërtius (on sects) | Megara |
| 3 | ps.-Plutarch, *Life of Plato* | Confused |
| 4 | Heron, *Geodaesia* | Mathematician |
| 5 | Proclus, *Commentary* | Mathematician |
| 6 | Marinus, *Data* | Mathematician |
| 7 | Aulus Gellius | Megara |

**Texts in Zamberti's collection**

Zamberti begins by quoting the entry on Euclid of Megara from the Byzantine lexicon known as the *Suda*. This turns out to be little more than a digest of the second text he quotes, the life of Euclid in the ancient *Lives of the Philosophers* by Diogenes Laërtius (2a). Diogenes's *Life of Euclid*, as we shall see, is the ancient source that most closely resembles a conventional biography, and it was to have an enormous influence on later investigations. According to Diogenes, Euclid was a native of Megara, between Attica and the Peloponnese. At first a disciple of Parmenides, he gathered a small following of his own, known as the "Megarics." Although his school did profess some positive doctrines ("the good is one thing, called by many names"), they were better known for their argumentativeness, which even drew the attention of comic poets. One mocked "wrangling Euclid, who inspired the Megarics with a frenzied love of controversy." This Euclid specialized in subverting logical argument altogether, by attacking only the conclusions of arguments, not their premises.

Diogenes went on to say that despite his combative nature, the philosopher Euclid enjoyed a close friendship with Plato; in fact, after the death of Socrates, Plato and the rest of the Socratic circle fled Athens to take sanctuary with him in Megara. Moreover, Euclid – like Plato and so many of the other followers of Socrates – devoted himself to writing Socratic dialogues with titles like *Crito* and *Alcibiades*; in all, Diogenes listed six of these dialogues. They were the only writings that Diogenes attributed to Euclid.

Zamberti then quoted two further extracts from Diogenes: first, a passage (2b) from his life of Socrates which listed Euclid among the followers of Socrates (along with Plato, Aristippus, and others); and second, a brief note on the sects of the philosophers, where Diogenes called Euclid the leader of the Megaric sect (2c). At this point, despite his promise not to interpose his own opinion, Zamberti broke in with a puzzled observation:

> It really is remarkable that this author makes no mention here of any of the other works which Euclid wrote.[11]

Zamberti's concern is understandable. There was no mathematical text more famous than the *Elements*; surely it should merit a mention in any life of its author, to say nothing of the many minor, but influential Euclidean treatises which also appeared in Zamberti's edition. It was certainly "remarkable" that a handful of obscure Socratic dialogues (none of them even extant) should receive sole billing. To put it more bluntly, there was nothing in Diogenes's biography to suggest that Euclid of Megara was a mathematician. In fact, with his taste for sophistical argumentation, Diogenes's Euclid seemed very unlikely to be the author of the *Elements*, a work of paradigmatic logical clarity and rigor.

Zamberti's suspicions may have been raised by Diogenes, but any temptation he may have felt to leap to a "rash judgement" regarding the identity of his author was assuaged by the next authority he quoted, the "*Life of Plato* by Plutarch" (3). According to this text,

> at the age of twenty-eight, Plato took the Socratic philosophers with him... and went off to Megara to visit Euclid, the most accomplished geometer of that time. Megara was a very prosperous city some twenty miles from Athens, where Euclid (a sometime disciple of Socrates) had his origin. When [Plato] had spent some time with him in intense study, he set out for Cyrene.[12]

---

[11] *Ibid.*, fol. 7r: "Mirum siquidem fuerit quod is auctor nullam aliorum operum ab Euclide conscriptorum fecerit mentionem."

[12] *Ibid.*: "Annos postmodum octo et viginti natus Plato Socraticis secum assumptis... ad Euclidem nobilissimum ea tempestate geometram Maegara secessit. Id autem oppidum florentissimum fuerat longe ab Athenis milia passuum viginti distans cuius oriundus erat Euclides Socratis aliquando discipulus. Cui cum aliquandiu studiosissime vacasset, Cyrenem profectus est."

This incident is clearly the same "flight to Megara" after the death of Socrates, which Diogenes Laërtius recorded in his biography. The author here explicitly identified Euclid the philosopher as a geometer, and even associated him with Plato's mathematical education.

This is much the same picture of Euclid that Valerius Maximus had provided, an author that Zamberti surely knew yet did not cite in his collection of sources. Perhaps he felt that Valerius's volume of morally improving anecdotes was not a serious historical source. But he had no reason to doubt the *Life of Plato* for, according to Zamberti, it came from the pen of none other than the philosopher and biographer Plutarch, whose lives of ancient Greeks and Romans were among the most highly regarded historical texts in Italian humanist circles.

Except for the fact that Plutarch never wrote a biography of Plato. The text that Zamberti quoted was actually a *Vita Platonis* written in 1430 by the Italian humanist Guarino Guarini of Verona. This text, freely adapted from the biography of Plato by Diogenes Laërtius, was printed many times in the fifteenth and sixteenth centuries as an appendix to Plutarch's *Lives* and it was no doubt its appearance in such a context that led Zamberti to think that it was a genuinely Plutarchan work.[13] The passage from which Zamberti drew his information went on to recount Plato's journeys to visit other mathematicians, such as Theodorus and the Pythagoreans, culminating in his journey to Egypt, where he learned the secrets of theology from the priests and became acquainted with the perennial philosophy which the Egyptians had borrowed from the Jews.[14]

The pseudo-Plutarchan *Life* assured Zamberti that Euclid the philosopher and friend of Plato was also the mathematician of the *Elements* – despite the peculiarities of Diogenes's biography and the further inconsistencies other texts would present to him. Read as a whole, Guarino's *Life* makes a persuasive case for a connection between Plato and the geometer Euclid, who appears as an instrumental figure in the mathematical turn which Plato took in his own philosophy after the death of Socrates. Guarino creatively merged Plato's "flight to Megara," an obscure detail from the philosopher's early life, with his famous journeys to the East in search of ancient wisdom. Taking refuge with Euclid after the execution of Socrates, Plato began a journey into the depths of mathematics; mathematics drew him on, via the Pythagoreans of Italy, to theology, which then led him to the truth hidden in the Jewish-inspired philosophy of the Egyptians. The conclusion to draw from Guarino's account was that Euclid's mathematics had been (and so continued to be) the essential propaedeutic to philosophical study – or, as the Renaissance humanists imagined, the first step in a philosophical journey to the Christian God.[15]

---

[13] See Sabbadini (1896, p. 136).

[14] Plutarch (1514, fol. 366v, obviously not the edition consulted by Zamberti).

[15] Ficino, in the life of Plato prefaced to his translation of the philosopher's works, seems to have been influenced by Guarino. Although Ficino does not mention Euclid, he presents Plato's trip to Megara as the first step of his travels in search of knowledge. Without mentioning the death of Socrates, Ficino says that, at about the age of 28 (which is how old Plato in fact was when his

Zamberti found this image of a philosophical Euclid particularly attractive. The preface he addressed to Guidobaldo, Duke of Urbino, at the start of his edition, was uncompromisingly Platonic, leading the duke through several pages of purely philosophical argument before mathematics was even mentioned. Starting with a quotation from Plutarch (this one genuine) on the endless disagreements among philosophers, Zamberti set out to settle these disputes once and for all by establishing the superiority of the Platonic philosophy over all other, materialistic philosophies.

The relevance of Zamberti's resolutely philosophical preface to the mathematical texts which follow becomes clear only after several pages of argument against materialism, and Epicurean atomism in particular. Zamberti was trying to show that mathematics – and above all Euclid's *Elements* – could provide solid arguments against the atomic theory. His first argument will illustrate his approach: the Epicureans maintain that an infinite number of atoms exist, moving through the infinite void. But the atoms must equal the infinite (that is, the total matter contained in all the infinite number of atoms must be infinite), and the infinite void is itself infinite. By Euclid's eighth common notion (the first in modern editions), things that are equal to the same thing are equal to each other. Thus the atoms (as a whole) must be equal to the void in which they exist, so that it would be impossible for them to move within it.[16]

As well as providing specific arguments against materialism, mathematics also played a more conventionally Platonic role in Zamberti's scheme. Citing Proclus's *Commentary* on Euclid, Zamberti located mathematics at a midpoint between the physical and divine. It served as a necessary bridge between the two, and a prerequisite for any philosophy which intended not just to refute materialism but to go beyond it.[17]

Guarino's *Life of Plato*, which identified Euclid of Megara with Euclid the mathematician, matched Zamberti's own convictions about the nature and utility of mathematics precisely. It provided a historical justification for the close relationship between Euclidean geometry and Platonic philosophy that Zamberti advocated in his preface. The cliché that mathematics was a preparation for philosophy was rendered vivid through the story of Plato beginning his own independent philosophical journey, after the death of Socrates, with the author of the *Elements*. Zamberti's misidentification of Guarino's text as an ancient, authoritative biography was quite opportune: of all the texts listed in his appendix, this imposter was the most important for his purposes, insofar as it was the only one that clearly demonstrated the connection between Euclid and Plato. Zamberti's certainty that the two men had been associates guided his choice of other texts and colored his interpretation of

master died), Plato travelled first to Megara, and thence to visit the places and mathematicians listed by Guarino. See Plato (1491, sig. a2r).

[16] Euclid (1505, fol. 2r).

[17] *Ibid.*, fol. 4r.

them. It also blinded him to the chronological impossibility of his identification of the two Euclids as one.

Zamberti presented several further testimonies for the life of "Euclid" (in fact, the two Euclids) in his 1505 edition. In his *Attic Nights*, Aulus Gellius recorded that in Socrates's time citizens of Megara were banned from Athens because of the animosity between the two cities. Euclid was so anxious to hear Socrates lecture that he crept into Athens by night, disguised as a woman. The story, which we encountered at the start of this chapter, reinforces the connection between Euclid and the followers of Socrates, even though it is evident to us that Gellius meant to refer to Euclid of Megara, not Euclid the mathematician, and nothing in his text indicates that he identified the two men. The other brief testimony is from Proclus's student Marinus, who records that the *Data* (a text included in Zamberti's volume) was indeed written by someone called "Euclid," perhaps relieving any doubts that Diogenes Laërtius's meager booklist had sowed in Zamberti's mind.

The longest of the final testimonies Zamberti cited was an extract from Proclus's *Commentary* on the first book of Euclid's *Elements*. Zamberti was one of the very first humanist scholars to be aware of Proclus's as yet unedited and unpublished *Commentary*, and in his preface, he quoted liberally from it, often without acknowledgement. But he seems not to have grasped what the text had to say about the identity of Euclid. Proclus provided Zamberti with the most precise chronological data on Euclid the mathematician then available – but this was information he nevertheless managed to misconstrue.

At the end of his historical survey of Greek mathematics (the "Eudemean summary"),[18] Proclus wrote that Euclid took his material from Eudoxus and Theaetetus (both mathematicians of the Academy), and imposed on it a rigor and order it had not had before. This Euclid lived during the reign of the Hellenistic King Ptolemy I of Egypt, with whom he had a famous exchange: asked by the king whether there was some easier way to learn mathematics, Euclid replied there was no "royal road" to geometry. He lived before Archimedes, who mentioned him by name in his own work. Finally, Proclus notes (in keeping with his Platonic interpretation of Euclid throughout the *Commentary*) that Euclid was a member of the Platonic sect, and that his goal in composing the *Elements* was the construction of the five Platonic solids.

It should be noted that Proclus himself, writing in the fifth century A.D., had no direct evidence for the date of Euclid. He had to make an informed judgment, based on the meager data available.[19] By well-established tradition, the theory of proportion in book 5 of the *Elements* was known to have been developed by Eudoxus, while the material on irrationals in book 10 was certainly by Theaetetus. These two companions of Plato together provided a *terminus post quem* for Euclid. The fact that Archimedes cited him established a *terminus ante quem*. Using this narrow chronological window, Proclus inferred that Euclid's famous comment that there

---

[18] See p. 7 above.

[19] The third edition of the *Oxford Classical Dictionary* (1996) states that Euclid's dates are uncertain – between 325 and 250 B.C. – and that it is only Proclus's "worthless inferences" that link him to Alexandria and the Hellenistic king Ptolemy I. In fact, "nothing is known of his life."

was no "royal road" to geometry must have been addressed to Ptolemy I. As he put it, "[Euclid] is younger than the followers of Plato, but older than Eratosthenes and Archimedes."[20] It was on the basis of this historical inference that Proclus appended Euclid to the end of the list of geometers recorded by Eudemus – a list which had ended with the mathematicians of the early Academy – saying that Euclid was "not much younger than these men." (The fact that Eudemus, a student of Aristotle, does not mention Euclid in his history of geometry provided Proclus with another *terminus post quem* – hence he could be sure that the contemporary geometers that Eudemus *did* mention predate Euclid.)

By claiming that the mathematician had compiled the *Elements* so as to construct the regular "Platonic" solids that appear in the *Timaeus*, Euclid could be connected to the earlier school of Plato that Eudemus had described. Such an interpretation of the *Elements* has no historical justification whatever, reflecting only Proclus's own Platonic prejudices and the Neoplatonic tendency to ascribe a single *skopos* or intent to authoritative texts.[21]

Proclus was sure that Euclid was a follower of Plato; nevertheless, the Eudemean summary provided strong evidence that Euclid the mathematician was not a *contemporary* of Plato's, but must have been considerably younger than him, since he was younger than Plato's followers and the geometers who succeeded him at the Academy. Proclus himself certainly did not confuse Euclid the mathematician with Euclid, the friend of Plato, however much he wished to cast the history of mathematics in a Platonic mold. A careful reader of Proclus's text could conclude that the lives of Plato and Euclid the geometer probably did not even overlap.

Yet this was *not* the conclusion that Zamberti reached. In fact, in his translation he subtly misrepresented Proclus's meaning, so that the philosopher's fairly unambiguous location of Euclid in the generation after Plato (or even later) was completely lost. Where Proclus had written that Euclid was "younger than the followers of Plato," Zamberti claimed instead:

> [Euclid] was not much younger than Plato, but lived a little later than his time. But he was older than Eratosthenes and Archimedes.[22]

This is not a translation at all; rather, Zamberti combined Proclus's clear statement that Euclid was younger than the followers of Plato, with another chronological statement that Euclid was "not much younger than these men" – that is, than, the geometers listed by Eudemus, Proclus's main pre-Euclidean historical source. Then, by comparing Euclid to Plato himself rather than to his followers, Zamberti comes up with a chronological statement that exists nowhere in the text he was supposedly translating: that Euclid was *not much younger than Plato*.

---

[20] Proclus (1873, p. 68): "*neôteros men oun esti tôn peri Platôna, presbuteros de Eratosthenous kai Arkhimêdous.*"

[21] Ramus critiques this view at some length, at Ramus (1569, pp. 43–44).

[22] Euclid (1505, fols 7r–v): "Non admodum iunior sed aliquanto posterior quam Platonis tempore vixerunt [*sic* for 'vixerit']. Sed Eratosthene et Archimede antiquior."

Zamberti made Euclid a younger contemporary of Plato by massaging the very text that showed that he was not. Nevertheless, he was aware that Proclus had made Euclid a little later than many of the followers of Plato; in the preceding paragraph, he had translated this statement quite accurately. In order to reconcile these contradictory statements, Zamberti added a phrase entirely of his own invention: that Euclid "lived a little later than his [Plato's] time." The reader is left with the impression that Euclid was a young man when Plato was in his prime, surviving him and even his followers by some years. This is no doubt how Zamberti himself imagined the historical situation, but his evidence was based on a wishful paraphrase (or even distortion) of Proclus's text. Whether he did it intentionally or not, Zamberti managed to make his most explicit genuine source on the date and identity of Euclid read quite differently from its author's intent.

To complicate the tale still further, Zamberti also took into account the meeting between Euclid and Ptolemy I that Proclus records. He used this *datum* to establish a very precise date for Euclid which entirely undermined his identification of him with Euclid of Megara. Having consulted "books of chronicles," he stated that Ptolemy I lived "4908 years after the creation of the world, or 291 years before the coming of the Savior."[23] This led him straight into chronological impossibility: "An auditor of Socrates, he lived in the age of Plato, during the reign of Ptolemy I."[24] Either Zamberti did not notice that someone alive in 291 B.C. would be an unlikely companion of Plato (d. 347), still less a follower of Socrates (d. 399); or this was one of the "disagreements" he preferred not to resolve.

## The Spread of the Megaran Euclid

Zamberti deserves credit for unearthing so many texts – some quite obscure – concerning the life of "Euclid." But his expectations of Euclid's Platonism were so overwhelming that he was unable to make sense of the contradictions in the evidence, even as he set it out. Renaissance authors after Zamberti continued to conflate the two Euclids. Like Zamberti, most failed to see the chronological difficulties inherent in identifying the two men. In 1506, the Volterran humanist Raffaele Maffei mentioned, in a brief biographical notice on Plato, that the philosopher had studied

[23] *Ibid.*, fol. 7v: "Si ergo Euclidis tempore primi Ptolemaei Aegyptii regis ex libris Chronicis datur intelligi quot anni ab ipso Euclide usque ad nostra tempora fluxerunt. Ptolemaeus igitur primus Aegypti Rex fuit anno a mundi creatione MMMMDCCCCVIII, ante Salvatoris adventum annis CCLXXXXI. Quibus annis CCLXXXXI si addas Annos MDV qui a salute nostra hucusque fluxerunt fiunt anni MDCCLXXXXVI. Ab ipso igitur Euclide usque ad nostram ... aetatem effluxerunt Anni MDCCLXXXXVI. Haec sunt quae de ipso Euclide habere potuimus."

[24] *Ibid.*, fol. 5v. The entire brief biography of Euclid reads: "Euclides vero vir inquam ingenii praestantissimi, qui elementa in unum collegit. Multaque ab Eudoxo, multa a Theaeteto perfecit, et hinc et inde sumpta proclivius et planius quam qui ipsum praecesserunt demonstravit. Vixit Platonis tempestate Socratis auditor, temporibus primi Ptolemaei. Antiquior vero ut inquit Proclus Lycius Eratosthene et Archimede qui uno et eodem tempore vixerunt."

geometry with Euclid. Maffei was clearly drawing on Guarino's *Life of Plato*, perhaps via Zamberti. Maffei added that Plato extracted from geometry some of the deepest secrets of nature and God – again, drawing Euclid into a Platonic orbit in terms of chronology, intent and content.[25]

In his *Encomium geometriae*, Girolamo Cardano wrote that Euclid, the author of the *Elements*, *Optics*, *Catoptrics* and *Phaenomena* was a contemporary of Plato, who put him into the *Phaedo*.[26] Proclus's *Commentary* was one of Cardano's major sources in this speech, and he went on to summarize Proclus's list of mathematicians of the Academy.[27] Cardano was so certain that Euclid was a contemporary of Plato's that he prefaced his summary by saying that all of these mathematicians came "after Plato and *Euclid*" – thereby expressly contradicting Proclus's statement that *Euclid* came after the mathematicians of the Plato's school.[28]

A more sober author than Cardano, Konrad Gesner, also fell into the trap laid by Zamberti. In his entry for Euclid in his *Bibliotheca universalis*, Gesner seems to have had Zamberti's collection of biographical sources in front of him. At the start of the long entry, Gesner wrote, "The philosopher Euclid of Megara, from the city of Megara on the isthmus, founded the Megaric sect, which he also called dialectical or eristic (that is, argumentative). He was a disciple of Socrates."[29] Gesner went on to summarize the information on Euclid of Megara found in *Suda* and Diogenes Laërtius. Had he stopped there, he would have had a perfectly accurate biography of the Socratic philosopher; but he continued by presenting material from Proclus's commentary on Euclid, as well as information found in editions of the *Elements*, thereby showing beyond any doubt that he identified the philosopher with the mathematician.

Gesner prefaced his citation from Proclus by saying that the Platonist introduced his life of Euclid with a list of those mathematicians who preceded him – which Gesner goes on to paraphrase, beginning with Thales and ending with Philip, disciple of Plato (and mentioning Plato himself along the way). At the end, Gesner wrote, "Not much younger than those listed here was Euclid, who wrote the *Elements*, in which he included many of the writings of Eudoxus and completed many of Theaetetus's

---

[25] [maffei commentaria 1506] fol. 254r: "Omnium doctrinarum praesertim Geometriae studiosissimus fuit, in qua Euclidem et Architam Tarentinum Theodorumque Cyreneum audivit ex qua sane scientia Secreta quaedam investigasse in libris suis a paucis comprensa quae ad rerum naturam tum divinitatem pertinerent."

[26] Cardano (1663, vol. 4, p. 443): "Fuit, ut in Phaedone apparet, Platonis contemporaneus Euclides, cuius ut vetustissimi, clarissimi extant Elementorum tredecim libri, tum Phaenomena, Optici, Catoptrici."

[27] See p. 13 above for Cardano's *Encomium*.

[28] Cardano (1663, vol. 4, p. 443): "post Euclidem et Platonem, Cleodamus Thasius ..." Cardano also wrote briefly about Euclid in his *De subtilitate*, stating that the geometer "sprang from Megara" ([cardano subtilitate 1560], p. 1011 (book 16): "Megara fuit oriundus.")

[29] Gesner (1545, fol. 226r): "Euclides Megarensis philosophus, ex Megaris urbe in Isthmo, sectam ab ipso Megaricam introduxit, quam et dialecticam et eristicam, id est contentiosam appelavit. Discipulus fuit Socratis."

works"[30] – an accurate rendition of Proclus's statement in his *Commentary*. It is quite extraordinary, then that Gesner puts the biographies from Diogenes (of Euclid of Megara) and Proclus side-by-side, without noticing the conflict – even though he was apparently aware that Proclus was listing the mathematicians who came *before* Euclid.[31]

A similar cognitive dissonance is to be found in Francesco Barozzi's 1560 Latin translation of Proclus's *Commentary*.[32] Barozzi rendered Proclus's statements about the life of Euclid quite accurately. In particular, the crucial sentences so mangled by Zamberti, he translated as: "Those who have written histories bring the completion of the science to this point. Not much younger than these is Euclid, who wrote the *Elements*..."[33] Thus he recorded unambiguously that Euclid flourished later than the generation after Plato. Nevertheless, Barozzi himself seems to have believed that Euclid of Megara wrote the *Elements* – despite being the first to publish in full, and in Latin the evidence that showed he did not.[34]

The difficulty with Proclus's chronology may have been that it was a scholarly *argument* rather than a primary, authoritative statement. Proclus triangulated a historical position for Euclid based on the fact that he did not appear in Eudemus's history of geometry but was cited by Archimedes and Eratosthenes. Readers who were already sure that they knew when Euclid lived – at the same time as Plato – seemed to have hurried through Proclus's somewhat complicated argument without even realizing that Proclus was weighing up a matter that was in doubt. Zamberti is the only reader before Ramus to pay careful attention to the question – and even he ended up emending Proclus's text so as to avoid the chronological difficulties it threw up for the Megaran authorship of the *Elements*.

On the face of it, it seems extraordinary that so many learned humanists, with their strong commitment to historical research and textual criticism, should have failed to notice the glaring difficulties in the received biography. But, like Zamberti, they were influenced by a longstanding tradition which suggested that the two Euclids were one. The account of Valerius Maximus which set Euclid and Plato to work together on the doubling of the cube, the appearance of Euclid as an interlocutor in Plato's dialogues, and the attribution of the *Elements* to Euclid of Megara in manuscripts and early printed editions each contributed something to the

---

[30] *Ibid.*: "Istis hactenus enumeratis non multo iunior Euclides, elementa conscripsit, quibus multa Eudoxi scripta comprehendit, et multa Theaeteti absoluit."

[31] Bernardino Baldi, in his life of Euclid written in the late 1580s or early 1590s, correctly distinguished between the two Euclids, and identified Gesner and Cardano as the principal sources of the confusion among his contemporaries. See Pace (1993, footnote on pp. 202–204). Gesner used a "cut-and-paste" method (quite literally) to assemble his monumental works, and that may account for the unassimilated juxtaposition of these texts. See Blair (2003, pp. 16–17).

[32] For a fuller treatment of this work, see p. 167.

[33] Proclus (1560, p. 39): "Qui itaque historias perscripsere, hucusque scientiae huius perfectionem producunt. Non multo autem his iunior Euclides est, qui Elementa collegit..."

[34] Proclus (1560, sig. *4r). He says that his path to translating Proclus began with his difficulties in studying "Euclidem Megarensem insignem mathematicum."

assumption. In addition, Zamberti himself had provided a new selection of ancient texts (or purportedly ancient ones) illustrating the life of Euclid, all of which seemed further to support the traditional identification.

The issue of authority must have played a part as well. Euclid, the author of the *Elements*, was a very shadowy figure compared to, say, Archimedes. But his text held an exalted place in the university curriculum. To justify the esteem in which he was held, Italian humanists needed its author to have a life, a school, a philosophical pedigree. Euclid was the only author whose name was synonymous with one of his works: to say, simply, "Euclid" was to invoke the *Elements* (not the *Optics* or the *Phaenomena*); and no other author could stand in so completely for his own writings. As one scholar has put it, in Euclid "the work and the man are the same."[35] It may seem slightly absurd to seek a grounding for Euclid's *Elements* outside the text itself; after all, surely Euclid's demonstrations would suffice to establish his authority. But, on the one hand (as we shall see in the next chapter) many scholars in the sixteenth-century did not believe that Euclid had written the demonstrations in the *Elements* – and often taught their students geometry without referring to the proofs at all. And, on the other, as editors of Euclid's text sought to broaden its appeal beyond the classroom, even to craftsmen and laborers, the text had to have something to recommend it to its readers besides its self-sufficient truth; to draw unlikely or reluctant readers into the *Elements*, its author had to carry some weight, and to be attractive even to those who had not yet read his work.[36]

In short, humanists needed Euclid to be the sort of authority, and to have the sort of intellectual pedigree that their fellow humanists would recognize. There had to be an appropriate biography of Euclid. By a happy coincidence, Diogenes's life of Euclid of Megara fit the bill. The peculiar fact that Diogenes's Euclid seemed not to have engaged in mathematics at all could be corrected with other texts, such as those that Zamberti edited in his preface.

Behind all this lay the impulse of a strongly Platonist apologetic, which Zamberti shared with many mathematical writers of the Renaissance. The dialogues of Plato had been recovered only recently in the Latin West. Assimilating Plato's philosophy and reconciling it with that of Aristotle were central philosophical concerns of the learned humanism of the day. It was clear that mathematics was central to Plato's theories of knowledge and being. Therefore, it seemed highly appropriate that Euclid the geometer should be his friend and colleague. Proclus himself had nearly said as much, when he wrote that Euclid was a Platonist and wrote the *Elements* to the end of constructing the five regular solids of the *Timaeus*. And there was the explicit (though erroneous) testimony of Valerius Maximus on Euclid's role in the doubling of the cube. Even if the philosopher who emerged from Diogenes's life was hardly a Platonist, this could be compensated for. In his *Life of Plato*, Guarino Guarini conflated Diogenes's account of the Socratics' flight to Megara with the story of Plato's journey to the East in search of ancient wisdom. Euclid

[35] Billingsley (1993, p. 3).

[36] This point made in Billingsley (1993), *passim*.

of Megara was invoked to explain Plato's mathematical turn, and his own philosophy (as recorded by Diogenes), which was so different from Platonism, was silently passed over. Thus, in a quite literal and historical sense, mathematics was made to occupy its accustomed place as a propaedeutic to philosophy.

## Ramus on the Biography of Euclid

A new chapter in the fortunes of Euclid's life opened with Ramus's *Prooemium mathematicum*. As Ramus approached the biography of Euclid, he confronted a complicated legacy of texts. There was Zamberti's confused and confusing melange of anecdotes, packed with chronological impossibilities. These had provoked only the slightest discomfort in Zamberti himself or in most of his readers, and the one text in the collection which might have given the game away – Proclus's *Commentary* – Zamberti had mistranslated in such a way that it seemed to offer no challenge to the Megaran attribution. Further supporting the traditional biography was a well-known anecdote from Valerius Maximus in which Plato was seen to defer to Euclid's mathematical knowledge. On the other hand, both the Greek text of Proclus's *Commentary* and Barozzi's excellent translation of it had become available since Zamberti. They both transmitted the crucial chronological passage in Proclus quite accurately; yet neither the editor of the text nor its translator noticed that Proclus's data made the traditional attribution of the *Elements* impossible. Publication of the relevant text was not enough to solve the problem. The solution would also require a reader capable of resisting the reflexive identification of Euclid the geometer with Euclid of Megara – a reader, in other words, who was predisposed, perhaps even determined, to put some distance between Euclid and Plato.

Ramus was precisely that reader. As he came to consider the biography of Euclid, he had several reasons to look at Proclus's dates with a fresh eye. Ramus developed an extended argument, or rather, a series of arguments sustained simultaneously through the first book of the *Prooemium mathematicum* and continued into subsequent books, challenging Proclus's historical model for the development of mathematics. Ramus's life of Euclid (which appears halfway through this book) formed a link in this chain of argumentation; his willingness to reconsider Euclid's place in the history of philosophy and mathematics is intelligible only within the context of this larger concern.

As we have seen, in the first book of the *Prooemium mathematicum*, Ramus recounted the history of mathematics from the antediluvian patriarchs to Theon, who (for reasons that will be explored in the next chapter) occupied the end point of ancient mathematics in Ramus's scheme. The larger part of the book consisted of an extensive biographical history of Greek mathematics, and the life of Euclid began immediately after his account of mathematics in the works of Aristotle and among the philosophers of the Lyceum.[37] Ramus maintained that Aristotle's philosophy

[37] The life of Euclid is on pp. 22–24 of the *Scholae mathematicae*.

was only comprehensible to one who had a solid grounding in mathematics, especially geometry. This was a barb thrown at his "ungeometrical" Peripatetic opponent Charpentier – one of many in the *Scholae*. But he was also trying to make an important point about the historiography of mathematics. Proclus drew the material for his rapid survey of the history of mathematics from the works of Theophrastus and Eudemus, who were students of Aristotle. But he did not mention these Peripatetic sources by name, nor did he offer any discussion of Aristotle or the other mathematicians of the Lyceum who must have figured in their histories. Instead, he limited his list of famous mathematicians to members of the Academy. Proclus had not so much borrowed from his sources as amputated (*detruncarat*) them.[38]

Part of Ramus's continuing complaint against Proclus, then, was that his history was partisan. He did not question its *accuracy* in the strictest sense – after all, it was drawn from dependable, early sources. Rather, it was its biased selection, and subsequent aggrandizement of the Platonic school, to which Ramus objected. In other words, as he worked up to his treatment of the life of Euclid, Ramus was already deeply invested in an effort to show that mathematics had been done – indeed, had thrived – outside of the Academy. He also tackled another claim made by Proclus (or at least, a claim which Ramus believed Proclus had made): namely, that the school of mathematicians from Plato to Philippus of Mendes (the last mathematician listed in Proclus's chronology before Euclid) had brought the art to a state of perfection:

> Proclus says that all these Platonists lived together in the Academy, and exercised each other with the questions they shared, and brought the mathematical philosophy to perfection. And so, in Proclus's judgement, Plato's academy invented mathematics, or encouraged it, or certainly perfected it.[39]

He repeated the charge on the following page: Proclus said that "mathematics in numbers and figures was discovered and perfected in this time."[40] In Ramus's view, this was nothing more than partisan hyperbole. He would show that many men outside the Academy had contributed to the development of mathematics.

The issue was of enormous importance for Ramus. In his view, the greatest mathematicians of antiquity were Archimedes and Heron of Alexandria, who certainly fell outside the golden age during which (as Proclus had it) mathematics had reached perfectio. If Proclus's statement were allowed to stand, the only conclusion would be that Archimedes and Heron did not come up with any *new* mathematics, just new *applications* of already-discovered truths – a conclusion that would strike yet another blow to the status of the applied mathematician. What is quite odd, however, is that Proclus never actually argued for such a mathematical "golden age." In his account of the Platonic mathematicians, he simply said that they "lived together in

---

38 Ramus (1569, p. 22).

39 Ramus (1569 p. 20): "Atque omnes hi Platonici in Academia unâ conversati, et communibus inter se quaestionibus exercitati, mathematicam philosophiam ad perfectionem deduxerunt, ait Proclus. Ita Procli judicio Platonis Academia mathematum inventrix, vel altrix, certe perfectrix efficitur."

40 *Ibid.*, p. 21: "Atqui periodus ista est, in qua putat Proclus mathematicam in numeris et figuris inventam et perfectam fuisse."

the Academy, making their inquiries in common."[41] This is the passage to which Ramus referred; but Proclus did not go on to say, as Ramus said he did, that these same philosophers perfected mathematics. The closest he came to such a statement was a remark to the effect that Amyclas, Menaechmus and Dinostratus made "the whole of geometry still more perfect."[42] Ramus translated this quite correctly on p. 19 of the *Scholae*.[43] It is just possible that Ramus misremembered this statement about the relative perfection of geometry, as if Proclus had applied it to the entire period in absolute terms. On the other hand, Ramus may deliberately have set up Proclus as an historiographical straw man.

Whatever the source of his error, when Ramus turned to the life of Euclid, he had Proclus's supposed era of mathematical perfection in his sights. Ramus started the life by reminding his readers:

> We remember that, according to Proclus's opinion, mathematics had already been perfected in the Academy. However, many things were discovered by those who came later. Four elementators have been listed so far: Hippocrates, Leo, Theudius and Hermotimus. Proclus makes the fifth elementator, Euclid, a little younger than the followers of Plato just discussed, and says that he lived under Ptolemy I and was known to him in Egypt. Valerius, on the other hand, says in 8.13 that when the keepers of the sacred altar tried to discuss its size and shape with Plato, he ordered them to consult Euclid the geometer, yielding to his knowledge – indeed to his profession.[44]

By juxtaposing an accurate reading of Proclus's crucial chronological statement with the testimony from Valerius Maximus, Ramus made the contradiction between the two sources evident – a contradiction he proposed to resolve by rejecting Valerius altogether:

> But Proclus, the master of geometry, had learned the truth about this matter from Theophrastus and Eudemus... and his version of this history seems to me more plausible than that of Valerius. I do not find anything on the doubling of the cube attributed to Euclid, but the whole thing is attributed to the prince of philosophers, Plato. *Nor did Euclid in the* Elements *say anything explicitly about doubling the cube; if this [i.e., Valerius's report] were really true, he would not have kept silent about it.* And so I accept from Proclus that Euclid was younger than Plato and his followers.[45]

---

[41] Proclus (1992, p. 56).

[42] *Ibid.*

[43] "Amyclas ... et Menechmus ... et frater ipsius Dinostratus longe perfectiorem Geometriam reddiderunt."

[44] Ramus (1569, p. 23): "Meminerimus igitur mathematicam adhuc e Procli sententia perfectam in Academia fuisse, cum tamen plurima deinceps a posteris inventa sint. *Stoikheiôtai* quatuor adhuc expositi sunt Hippocrates, Leo, Theudius, Hermotimus. Quintus Euclides deinde a Proclo Platonicis commemoratis paulo iunior efficitur, et dicitur sub Ptolemaeo primo floruisse, eique etiam notus fuisse in Aegypto. Valerius tamen ait libri octavi capite decimo tertio conductores sacrae arae modum et formam eius cum Platone conferre conatos, ad Euclidem Geometram ire iussos, scientiae eius cedente, immo professioni."

[45] Ramus (1569, p. 23): "Sed Proclus Geometriae magister a Theophrasto atque Eudemo praesertim hac de re veritatem edoctus... in hac historia mihi verisimilior est, quam Valerius: nec duplicati cubi quicquam ad Euclidem, sed ad Platonem principem totum referri comperio. *Nec Euclides in*

Here, then, the Euclid problem was finally resolved. Ramus understood the meaning and significance of Proclus's chronology, and brought out the clear contradiction between Proclus's late Euclid and the contemporary of Plato whom Valerius celebrated.

Confronting this contradiction, Ramus said he had several good reasons for following Proclus rather than Valerius. First, Proclus, partial though he may have been, nevertheless drew on excellent sources, the histories of geometry by Theophrastus and Eudemus. Second, Valerius was the only authority to say that Plato turned to Euclid for help with the duplication of the cube; other sources attribute the "whole thing" to Plato. Finally, in a passage he added for the *Scholae*, Ramus wrote that if Euclid *had* found a way to double the cube, he would have mentioned it in the *Elements*.

This reasoning needs to be examined more closely. Ramus's strongest argument for trusting Proclus over Valerius was, surely, that Proclus drew on early and hence more reliable sources. But this argument is undermined by a rather obvious chronological error: Proclus may have relied on Eudemus and Theophrastus for some information in his *Commentary*, but these two authors lived *before* Euclid and so could not have served as sources for Proclus's life of the geometer. Indeed, Proclus made it very clear that his historical sources gave out just short of Euclid; as we have seen, he had to infer the date of Euclid indirectly.[46]

Ramus's second reason for preferring Proclus to Valerius was that Valerius's claim that Euclid had participated in the debate over doubling the cube was attested by no other ancient authority. Other sources, Ramus said, attributed the "whole thing" to Plato. But on this point, too, Ramus was wrong. In fact, *none* of the extant versions of the Delian anecdote attributed the solution of the problem to Plato alone: either he was said to have directed the questioners to other mathematicians of the Academy, or he was said merely to have interpreted the oracle as a reprimand against the Greeks for neglecting geometry (without devising a solution for it). Ramus, however, had come upon a list of ancient "solutions" of the cube duplication, in which Plato's name figured prominently. The Archimedean commentator Eutocius listed twelve (approximate, mechanical or non-planar) solutions to the problem,

> the first of which is Plato's, whose mesograph for finding immediately two mean proportionals is quite extraordinary; and so, at that time, Plato alone held the highest rank in mathematical accomplishment.[47]

*Elementis quicquam nominatim de duplicando cubo proposuit; alioqui tamen si haec vera essent, non taciturus* [this sentence was not in the 1567 *Prooemium*, added in the 1569 *Scholae*]. Itaque Euclidem Platone et Platonicis iuniorem a Proclo accipio."

[46] Proclus (1992, p. 56): "*All those who have written histories* bring to this point their account of the development of this science. Not long after these men came Euclid ..."

[47] Ramus (1569, p. 15): "Quarum [sententiarum] prima et ingeniossima est Platonis, cujusque mesographus ad duas medias protinus inveniendum, singularis est: Mathematicae itaque laudis principatus tum penes unum Platonem fuit."

Ramus used this reference to the philosopher's mesograph to justify his insistence that Plato had solved the Delian problem entirely on his own.[48]

He also had another source, one which sheds further light on his redating of Euclid. I discussed this source, Philoponus's *Commentary* on the *Posterior Analytics*, in a previous chapter.[49] I showed there that Ramus related this anecdote as the central episode in his life of Plato. As he told the story in the *Prooemium*, Plato had alerted the learned world to this problem, adopting the same respected role at the center of a network of mathematicians that Ramus believed himself to occupy. When Ramus returned to the Delian anecdote in his life of Euclid just a few pages later, the story continued to carry the weight of analogy between himself and Plato. It was significant that he returned to this particular anecdote. He could, after all, have chosen *any* of the incidents from the life of "Euclid" (really Euclid of Megara) – his disguised visits to Athens to hear Socrates, the sanctuary and teaching he provided Plato after the death of their master – and shown that it was inconsistent with Proclus's statement that Euclid lived later than the followers of Plato. A student of Socrates during the conflict with Megara, for instance, could not still be alive a generation or more after Plato. But each time Ramus invoked the Delian problem in the *Scholae mathematicae*, it was to identify himself with Plato: the philosopher who castigated the ungeometrical Greeks and who (in the detail invented by Ramus) wrote letters throughout the known world, bringing his friends to their mathematical senses.

In Ramus's earlier retellings of this anecdote, Plato was made to appear as a critic, mathematical intelligencer and leader of a school: precisely the roles that Ramus performed in the *Scholae mathematicae*. And, the final time it was evoked – when Ramus considered Valerius Maximus's version of the story – Plato was still acting as an *alter Ramus*. By insisting that Euclid did not belong in this story, Ramus was putting further distance between *himself* and the author of the *Elements*. With his 1555 *Arithmetica*, Ramus had rejected his early enthusiasm for Euclid; the *Elements* was no longer the vehicle for mathematical instruction, but the problem that mathematical reform had to overcome. It was inconceivable that Ramus, writer of his own mathematical textbook, international correspondent and restorer of mathematics to the University of Paris, should turn to *Euclid* of all people when confronted by a difficult problem. Ramus went further even than his earlier retellings of the story when he wrote that "the whole thing was done by Plato ..." – as if to underline that the whole work of geometry should be attributed to *Ramus*, not to Euclid.

The key point here is that Ramus more or less stumbled on the correct answer, not in a spirit of disinterested historical inquiry, but in pursuit of a polemical point. It was not that he discovered, in Philoponus, that Euclid had nothing to do with the doubling of the cube and *therefore* concluded he must have lived later. Rather, Ramus was *determined* that Euclid should have nothing to do with the doubling

[48] A description of the mechanical solution that Eutocius attributed to Plato can be found in Knorr (1986, pp. 57–61), which shows that the association of such a mechanical contrivance with Plato is very dubious; Eratosthenes was most likely the inventor.

[49] See text cited in n. 26 of third chapter above.

of the cube, and *therefore* found historical grounds for assigning him a later date. That later date also supported the larger argument that Ramus had been pursuing all along, namely that mathematics had continued to develop outside and beyond the Academy of Plato.

Put another way, it seems to me that Ramus could not have considered a late Euclid unless he had thought that a late Euclid made historical sense. The idea of an early, philosophical Euclid, a contemporary and companion of Plato, had made so much sense to Platonist-leaning humanists that they could not seriously question it. Now, this early Euclid made such *little* sense to Ramus that he was driven to question the evidence for it.

There was one final reason why Ramus was willing to reconsider the dating of Euclid: his new historical model required some time to elapse between Plato and the author of the *Elements*. The *Prooemium mathematicum* marks a real departure from the historical schema Ramus had developed in his various earlier prefaces. In the preface to his 1555 *Arithmetica*, Ramus first set out his case against the *Elements*, and there he laid the blame for the later decline of mathematics squarely at the feet of Euclid and Theon. In the history of mathematics, they played much the same role that Aristotle did in his history of philosophy: vain, unprincipled individuals who perverted the simplicity of their art in order both to claim originality and to limit knowledge of the art to those who declared themselves disciples. As I noted in a previous chapter, Ramus's attack on the personalities of Euclid and Theon had little historical motivation; his anger with Euclid in particular seems to have arisen from his frustration with the *Elements* as a beginner's text in geometry.[50] By the time he wrote the *Scholae mathematicae*, Ramus had developed a more sophisticated historical model, in which the corruption of mathematics was to be attributed not to a particular individual, or group of individuals, but to the very idea of an abstract mathematics without applications. This idea constantly tempted even the best mathematicians; it promised elitism, the power of having mathematics in one's own control and elevation to the divine. Plato, despite excelling in mathematics, had embraced the error, passing it on with such force that even the great Archimedes could not wholly resist it. Ramus's history of mathematics thus became, in part, the account of a dangerous idea and its perennial appeal. It is an account that carefully separates the mathematical achievements of historical actors from their influence on posterity as teachers or theorists of the nature of mathematics. In making this distinction, Ramus imitated the historical model he had developed for rhetoric, in which the consummate practitioner of the art (Cicero) also exerted the most baleful theoretical influence.

In the new narrative of the *Scholae mathematicae*, Plato had a crucial role to play. As I showed in a previous chapter, Ramus saw the history of mathematics up to Plato as a heroic age. It came to a great climax with Plato himself, who reunited several divergent traditions and was the "prince of mathematicians."[51] He made

---

[50] See text cited at n. 49 of second chapter, and discussion at p. 35 above.

[51] See p. 43 above.

several important discoveries, and also perfected the analytical method of mathematics (as Ramus understood "analysis"). Just as he was the father of the one dialectical method, so too did Plato perfect the only correct mathematical method, the method that Ramus would himself use in his *Geometria*. But, as we have seen, Plato occupied the highest station in the mathematical firmament only to rain down destruction on the art over which he presided. Hints of corruption had appeared before him, as others yielded rarely to the destructive lure of abstraction: the use of demonstration, and the introduction of the reduction to absurdity, in particular, were signs of its influence. When Plato took mathematics out of the hands of practical men and teachers, reserving it for a philosophical elite, he signalled to his followers that mathematics should henceforth be wrapped in obscurity, and thus protected from the unworthy. Unnatural ordering and demonstration, which until then had been aberrations, became the normal way of teaching and presenting mathematics.

Euclid's *Elements* represented to Ramus the nadir of corruption. As he demonstrated at length in Books 3 and 6–31 of the *Scholae mathematicae*, the original pure simplicity of mathematics had been entirely obscured by perverse disorderings and the now ubiquitous demonstrations. But, if the Megaran authorship stood, Ramus would be at a loss to explain how Plato, the prime mover of the corruption of mathematics, and Euclid, its final product, could have been contemporaries. The decline of mathematics needed some time in which to take place. Thus Ramus was moved to reconsider the era of Euclid's, just as he would extend the composition of the *Elements* all the way to Theon, the last ancient mathematician.[52] Only in this way could there be sufficient time for the forces of corruption and decline to do their work.

One last consequence of the new historical model Ramus worked out in the *Scholae mathematicae* was that it allowed Ramus to present the *Elements* as the result of a historical struggle between, on the one hand, the discovery (or rediscovery) of mathematical truths over historical time and, on the other, the disastrous rise of abstraction, leading always to obscurity, elitism and eventually the neglect of mathematics itself. Convinced of the truth of this model, Ramus admired the early elementators, even though he had little biographical information to go on. This did not stop him from expanding imaginatively upon Proclus whenever he made even the briefest remarks about their discoveries. From Ramus's point of view, it would be manifestly unfair for Euclid to have more of a biography than his predecessors, when he (and Theon after him) had ruined what they had so much more wisely built.

Moreover, the association between Euclid and Plato was part of the humanist elevation of Euclid himself into a mathematical genius. For this reason Ramus had to reject the story in Valerius Maximus and the entire notion of a Platonic Euclid. In any case, to credit him with duplicating the cube would be to allow him too concrete an achievement. And Ramus went even further than this. With an argument that will be examined in the following chapter, Ramus claimed that Theon, the last of

[52] See discussion at p. 170.

the elementators, was more responsible for the present form of the *Elements* than Euclid. As a coda to this argument, Ramus surveyed the other texts attributed to Euclid – the *Optics*, *Data* and *Phaenomena* – and noted that all had been attributed in the past to Theon as well. It was in this context that Ramus made his triumphant boast, "nothing is left to Euclid except an empty name."[53] From a writer whose authority derived from his solid Platonic pedigree and biography, Euclid became, in Ramus's hands, a mere cipher, his most famous work nothing more than the endpoint of a long, almost impersonal process of deterioration.

## Henry Savile's "Bold" Conjectures

When Savile wrote his lecture on the life of Plato, he included most of the stories that had become traditional among earlier Renaissance biographers – including the story that Plato studied mathematics with Euclid of Megara.[54] But in his life of Euclid, in the very next lecture, he no longer accepted the traditional account of the friendship between the philosopher and the mathematician.[55] Savile began his biography of Euclid with a survey of the ancient sources on the subject. He summarized the biographical details on Euclid of Megara from Diogenes Laërtius, the *Suda* and Aulus Gellius (perhaps using Zamberti's convenient sylloge at the front of his *Elements* of 1505).[56] Savile added that Euclid appeared frequently in Plato's dialogues as a proponent of the Megaran philosophy, characterised by dialectical and eristic logic-chopping. (Clearly he had not returned to the dialogues, for Plato mentions Euclid only in passing and reveals nothing of his philosophy).

Savile went on to relate the anecdote on doubling the cube from Valerius Maximus without comment. Then – like Ramus – he turned to Proclus's *Commentary* on the first book of the *Elements* and his story of the meeting between Euclid and the king of Egypt, Ptolemy I. Proclus concluded from this encounter, wrote Savile, that Euclid was a little younger than the *platonici*, and flourished in the reign of Ptolemy.[57]

---

[53] Ramus (1569, p. 39): "ut Euclidi praeter inane nomen nihil admodum relinquatur." See the discussion in third chapter, above.

[54] MS Savile 29, fol. 36v.

[55] MS Savile 29, fols 41r–44v.

[56] There is a long entry for Euclid in Savile's list of *auctores mathematici*. The entry begins (at MS Savile 28, fol. *8r) with a summary of Diogenes's life of Euclid of Megara, followed by details of published works by the mathematician Euclid. At the end of the entry, Savile notes that he flourished under Ptolemy I, that he is mentioned by Aulus Gellius, and that Zamberti has information on him taken from Proclus. Five lines of the entry consist of disjointed notes taken from Ramus's *Prooemium mathematicum*. When he compiled this entry (as when he wrote his lecture on Plato), Savile clearly still thought that Euclid of Megara wrote the *Elements*, despite his marshalling of all the available sources on the biography of Euclid.

[57] MS Savile 29, fol. 41r.

Then, with a contrived display of humility, Savile revealed the conclusions he had reached from comparing these two sources:

> But now, my audience, I cannot hide my opinion. I shall not offer it as an oracle, nor as something that I shall feel obliged to defend later; instead, I shall offer it in such a way that it would be blameless to withdraw it if I do not convince you now. For I suspect that Euclid, the pupil of Socrates, the founder of the Megaran school who is mentioned in so many of Plato's dialogues, is a different man [from the author of the *Elements*], and of an earlier period.[58]

Savile went on to spell out the contradiction between Proclus's account and the common belief that Euclid the mathematician was Euclid of Megara. His first argument pinpoints the dates of Euclid of Megara and of the author of the Elements by means of two historical events:

> For between the Sicilian war, which was commanded by Euclid's fellow student Alcibiades (then older than Euclid, as was proper), and the death of Alexander, whom Ptolemy succeeded in Egypt, there were 93 years; yet if we believe Proclus, our Euclid was familiar with Ptolemy – and many years after the latter's elevation, as I believe, once he had established his rule and settled the affairs of the kingdom.[59]

The improbably long lifetime of Euclid, if he were one man, should be enough to convince anyone that these were two different men, living generations apart. Savile adduced several more arguments in support of his claim. Far from making Euclid the disciple of Socrates, he said, Proclus presented him as younger not only than Plato, another disciple of Socrates, but younger also than Eudoxus the disciple of Plato – and younger even than Menaechmus the disciple of Eudoxus. Proclus took some pains to show that Euclid was older, however, than Archimedes and his contemporary Eratosthenes, citing a passage from Archimedes where the mathematician explicitly mentions the *Elements*.

Savile was thus aware, as no earlier reader of this contentious passage had been, that Proclus was making a historical *argument*, inferring the date of Euclid from various circumstantial pieces of evidence – and he saw that this evidence was open to assessment and corroboration. The very fact that Proclus put forward an argument to show that Euclid lived before Archimedes was significant. For, Savile asked,

[58] *Ibid.*: "Atque hic, auditores, non possum dissimulare sententiam meam; quam profecto proferam non tanquam oraculum vel quod mihi necesse sit postea defendere – sed ita uti sit integrum revocare si eam hoc tempore vobis non probâro. Euclidem igitur illum Socratis discipulum, qui sectam megaricorum instituit, cuius toties mentio fit in libris Platonicis, suspicor alterum quendam et aetate superiorem."

[59] MS Savile 29, fol. 41r: "Nam inter bellum Siculum cui Alcibiades Euclidis condiscipulus iam aetate ut par erat provectior, praefuit, et mortem Alexandri cui Ptolemaeus in Aegypto successit, anni intercesserunt 93; et tamen hic noster Euclides, si Proclo credimus, cum Ptolemaeo familiariter versatus est, multis quidem, ut credo, post inaugurationem annis constitutis iam et pacatis negociis." For Alcibiades's command of the Sicilian expedition in 416 B.C., see Thucydides, *History*, VI.8; Diodorus Siculus, *Bibliotheca Historica*, XII.84; Plutarch, *Life of Nicias*, 14–15.

> is there anyone who does not see how pointless this argument would be, and how completely unnecessary, if Proclus was thinking of the Socratic philosopher mentioned so many times by Plato – who lived two centuries before Archimedes?[60]

In other words, Proclus may not have known for sure when Euclid the geometer lived, but he certainly knew that he was not Euclid of Megara.

It is clear, Savile concludes, that Diogenes Laërtius and the *Suda* were not referring to Euclid, the author of the *Elements*, at all, but rather to another person altogether: Euclid of Megara, the Socratic philosopher. This explained the eccentric choice of Euclidean texts which these authorities record: only philosophical dialogues, and nothing mathematical. Even if Euclid the geometer had written some dialogues in his youth, they would surely not be the only works his biographer would mention. Proclus, on the other hand, did not refer to any dialogues at all. He did not, in fact, connect Euclid with Socrates in any way, stating that he subscribed to the Platonic school, which was unconnected with, indeed antagonistic towards the eristic philosophy of Megara. This last point was a particularly astute observation, which had not been made by any preceding commentator; too many biographers of Euclid were all too ready to identify "Socratic" with "Platonic."

Savile's arguments were comprehensive, cogent and well-researched; this was the most complete resolution of the Euclidean question to appear in the sixteenth century. It was not quite as bold as Savile made out, of course, since the bones of it at least had been constructed by Ramus in the text Savile had read so closely, the *Prooemium mathematicum* of 1567. Indeed, it seems likely to me that, between his lectures on Plato and Euclid, Savile went back to reread the Euclid section of the *Prooemium* and was struck for the first time by the significance of Ramus's arguments.

Savile's silence about Ramus is not altogether surprising; after all, Ramus was his single most important source throughout the *Prooemium*, yet the one author he never cited. What is perhaps more surprising is the extent to which Savile *agreed* with Ramus. If Ramus had separated Euclid from Plato for his ideological, anti-Platonic purposes, shouldn't Savile have endeavored just as hard to associate the philosopher with the mathematician once again? But that would be to overlook the sheer historical interest of this problem. Savile's collection and assessment of the chronological data, together with his close reading of Proclus's argument, was humanistic scholarship at its best. Half a century later, when he delivered a series of lectures on Euclid to inaugurate the Savilian professorship in geometry, he picked out his identification of Euclid as the single achievement of the 1570 lectures that he remembered with most pride, and repeated most of the arguments from those youthful writings.[61]

---

60 MS Savile 29, fol. 41r: "Hoc argumentum quam esset inane minimeque necessarium si Socraticum illum intelligat toties a Platone, qui duabus seculis praecessit Archimedem, inductum, quis non videt?"

61 Savile (1621, p. 7): "In hanc sententiam de duplici Euclide, disputatum est a me ante annos quinquaginta et quod excurrit, cum in scholis publicis pro meo modulo interpretarer in ordinariis lectionibus Almagestum Ptolemaei."

## The Final Resolution

In 1572, two years after Savile's lecture,[62] Federico Commandino published a great edition of the *Elements* to which he prefaced a few biographical remarks. He noted that many had taken Diogenes Laërtius's life of Euclid of Megara to be that of the geometer, so that the author of the *Elements* was thought to be one of the Socratic philosophers. But, he observed, that could not be the case, since Proclus stated quite clearly that Euclid lived in the time of Ptolemy I: it would be impossible for the two to be the same person. In concluding remarks, Commandino revealed his source for his historical observation:

> But no one should think that I am unaware of Valerius Maximus' account: that Plato sent the keepers of the sacred altar to Euclid, as if he were the leading mathematician. But I follow those men most accomplished in the study of mathematics, Hero and Proclus, or rather Eudemus and Theophrastus, the greatest of the Peripatetics after the founder of that school.[63]

Commandino's reference to "Eudemus and Theophrastus" reveals that his source for the identity of Euclid was Ramus's *Prooemium mathematicum*, where the same two Peripatetic authors were erroneously claimed as Proclus's sources for the dating of Euclid. But Commandino's name carried an undisputed authority among both mathematicians and humanists – a stature to which Ramus had only been able to aspire. Commandino's separation of the two Euclids appeared, moreover, in a prominent place; by contrast, many readers of Ramus's *Prooemium mathematicum* seem to have missed the few lines correcting the record among the hundreds of pages of polemical history and invective.[64] With the publication of Commandino's

---

[62] In his *Praelectiones*, Savile drew attention to this fact rather caustically, saying that it was "only fair to believe" that Commandino had been moved by the same arguments as he had. (Savile 1621, p. 7: "In quam opinionem biennio postea Federicum Commandinum Italum iisdem, uti credere par est, permotum argumentis video incidisse.")

[63] Euclid (1572, sig. *5r): "Nemo autem mihi ignotum esse arbitretur, Valerium Maximum scribere Platonem sacrae arae conductores ad Euclidem, tanquam ad primarium mathematicum reiecisse. Sed nos Heronem et Proclum matheseos studio insignes sequimur, vel potius Eudemum ac Theophrastum ex peripateticis post praeceptorem nobilissimos."

[64] The problem of the two Euclids had been raised and resolved a couple of times before in print, but these solutions seem to have passed unnoticed. Caspar Peucer, in an oration delivered in 1557, raised the possibility that there were two Euclids of Megara, an earlier philosopher and a later mathematician (Melanchthon 1834–1860, vol. 12, col. 262). He had probably noticed the impossibility of the chronology; already in the table of mathematicians included in his 1553 *Elementa doctrinae* he had dated Plato to around 390 B.C., and Euclid of Megara to 292 B.C. – presumably intending by this the author of the *Elements* (Peucer 1553, sigs A4r–v). It is surprising that Savile, who used Peucer's book frequently to draw up his list of *auctores mathematici*, never noticed this feature of his chronology. In 1562, the Sicilian mathematician and humanist Francesco Maurolyco published a collection of sources on Sicilian history. He quoted a letter written by the fifteenth-century Byzantine historian Constantinus Lascaris to the prorex of Sicily, Fernando Acuña, concerning

correction, the Megaran error disappeared, as such obvious errors tend to do, into a kind of embarrassed silence. Euclid of Megara, the Platonic mathematician, simply ceased to exist.[65]

---

famous ancient Sicilians: "Euclid of Gela, the Platonic philosopher and most famous geometer, is a different man from the one whom Laertius writes about and who composed dialogues. As Proclus says in the second book of his commentary on the first book of Euclid (and as Heron also writes) he lived in the time of Ptolemy I, and was younger than Plato, but older than Eratosthenes and Archimedes. He was from Gela, as one can infer from the words of Laertius. He wrote thirteen books of Elements." (Maurolyco 1562, fol. 21r). Lascaris's intention (shared by Maurolyco) was to demonstrate that Euclid the geometer was in fact a native son of Sicily; Diogenes had reported that Alexander Polyhistor said that Euclid had been born in Gela, not Megara, and this was the clue that led Lascaris to consider the existence of two Euclids. See (Heiberg 1882), pp. 22–25 on Lascaris, Maurolyco and the *fortuna* of the Sicilian Euclid.

[65] In his sober *Bibliotheca philosophorum classicorum authorum chronologica* published 20 years later, Jean-Jacques Frisius has a biographical entry for Euclid of Megara under the year 422 B.C. (Frisius 1592, fol. 13v), and Euclid the geometer has an entry at 320 B.C. In neither entry is there any mention of the fact that the two men were once confused with each other. On Frisius, see sixth chapter.

# Chapter 6
# Rending Hypatia: The Body of the *Elements*

## Introduction

In the previous chapter, I considered the lengths to which Renaissance humanists went to construct a biography for that most familiar of ancient mathematicians, Euclid. As I argued there, the story of Euclid, the friend of Plato, was crafted in part to support a Platonic notion of mathematics, in part to provide a distinguished biography, one worthy of this most authoritative of ancient writers. Yet it was also an honorable attempt to make sense of suggestive, but contradictory evidence. However strong their convictions about the nature of mathematics, humanists writing the history of the art were of course constrained by what had actually survived from antiquity. The fragmentary state of the evidence and the scholarly errors they made in interpreting it led humanists down particular paths in mathematical history. Evidence about the mathematical past had survived haphazardly, at best; information about Euclid was very spare indeed. Humanists had to impose some interpretation onto this material. When they did, they were misled as much by textual contingencies and the happenstance of there being two men with the same name, as by their preexisting notions of the relationship between mathematics and philosophy, or their desire to elevate the person of Euclid.

The collapse of the Platonic "Euclid of Megara" resulted from much the same blend of ulterior motives and honest scholarship that had brought about the erroneous identification in the first place. Ramus, who was the principal debunker of the false Euclid, was motivated by his animus against the Platonists and their claim on mathematics, to be sure. In his zeal to rewrite the historical link between Plato and the geometry of the *Elements*, to reimagine both Plato and the development of the *Elements*, he would take away not only Euclid's identity as a friend of Plato, but (in the arguments considered in this chapter) even his agency in composing the *Elements*.

Nevertheless, Ramus's interpretation of the historical evidence for the life of Euclid was entirely correct: Euclid was younger than the followers of Plato, not a contemporary of the great philosopher. What is more, at a certain point, historical plausibility could trump even the most fervently held ideology. Ramus's critic Henry Savile, who wrote with a strong bias *towards* Platonism, embraced Ramus's

R. Goulding, *Defending Hypatia*, Archimedes 25,
DOI 10.1007/978-90-481-3542-4_6, © Springer Science+Business Media B.V. 2010

arguments against Megaran authorship, to the extent of claiming them as his own. Even though a Platonic Euclid would have suited his own agenda far better, Savile, like most humanists, could not resist the lure of solving a historical puzzle.

This chapter turns from the person of Euclid to the text he wrote, the *Elements*. Much of what I have just said about the biography of Euclid applies also to the problem of the text as a historical artefact. Scholars were motivated to criticize and historicize the text of the *Elements* by their philosophical presuppositions, or commitments to certain styles of pedagogy. But as this chapter will show, early textual criticism of the *Elements* was ensnared in a complex web of interconnected errors, even more intricate than that woven around the person of Euclid. At times, humanists misread or distorted the available evidence because of their preconceived notions, whether philosophical or pedagogical. More often, they failed to examine the evidence at all, accepting a convenient and supposedly well-established position. And as before, there was also a contingent element: the evidence itself was patchy and ambiguous, or was misinterpreted for entirely benign reasons. Some of these errors were eventually corrected by scholars who rejected the very enterprise of criticizing the *Elements*; scholars, that is, who had their own ulterior motives. Just as often, however, the errors were revealed simply through disinterested, even pedestrian, historical scholarship.

The problem treated in this chapter is, in a sense, a converse to the problem of identity of Euclid. There, scholars had conflated two men into the impossible "Euclid of Megara, author of the *Elements*." Once the historical absurdity became too evident to overlook, the two men were quietly separated from each other. Very little depended upon the identification besides a historical basis for the Platonic nature of mathematics – and (as I showed with the case of Savile) it was quite possible to defend that philosophical position in other ways, while accepting the new, more anonymous Euclid.

Here, by contrast, the problem is of a single text – the *Elements* – divided in two. For reasons that will be fully explored below, Renaissance scholars had come to believe that the *Elements* was the work of two authors, Euclid and Theon, the former responsible (only) for the statements of the propositions, and the latter for their demonstrations. Unlike the unification of the Euclids, this division of the *Elements* was more difficult to undo. First, the separation of the *Elements* into two sections of different value had consequences for the way in which the text was printed: it gave editors license to alter or add to the text quite drastically (at least those parts which were thought not to be Euclidean). Moreover, the debate over the status of the text became tangled up with another historical problem, itself bedevilled by even more serious misreadings of evidence and resulting historical absurdities: the dating of Proclus, whose commentary on the first book of the *Elements* provided the earliest witness to the state of Euclid's text.

There is another way in which the problem of this chapter differs from that of the last. The misidentification of Euclid – whatever its significance at the time for the philosophy of mathematics – was, in the end, just a strange historiographical glitch that was inevitably righted. The "Theonine hypothesis," on the other hand, opened up the very possibility of criticizing the *Elements* as a *text*, in that respect no different from Cicero's letters, the *Aeneid* or the New Testament – all texts that had been edited and historically criticized by Renaissance humanists. The Theonine hypothesis had extraordinary longevity; it was still current in the eighteenth century, and the underlying assumption – that Theon was a *bad* editor or commentator – was still alive until very recent scholarship.[1] Even as the historical errors were resolved, or their existence forgotten, the *possibility of criticism* that the Theonine hypothesis opened up could not be abandoned.

This chapter, then, will first examine the origins, and consequences, of the idea that Theon wrote the demonstrations to the *Elements* as a kind of commentary appended to Euclid's list of geometrical propositions. I then turn, once again, to the influential mathematical writings of Peter Ramus, showing how he added an extra dimension to this view of the *Elements* by bringing in the misconception that Proclus (a philosopher in fact of the fifth century) predated the fourth-century Theon by two centuries or more, and hence was a witness to an earlier, "pre-Theonine" recension of the text. Ramus used this evidence to dismember the *Elements* (a metaphor he employed himself, and which will prove to be significant in appreciating Savile's opposition to Ramus's project) into a jumble of earlier and later parts, thoroughly historicizing a text that, for medieval readers and even for Proclus himself, had stood altogether outside history, a repository of eternal mathematical truth.

Ramus's bold textual intervention raises a fascinating historiographical problem: how did he manage to place Proclus *three hundred years* outside of his true historical place? As I shall show, he was hardly alone in making this mistake. Modern scholarship has taken no account whatsoever of the Renaissance sense of the history of the Platonic school, which was plagued by confusions and systematic distortions. The magnitude of these historical errors, and their absence from modern accounts of Renaissance Platonism, justify devoting the central section of this chapter to unravelling these errors, their sources and eventual resolution.

Finally, I examine Savile's defense of the text of the *Elements* (and Euclid's single authorship), which appeared in his *Praelectiones* of 1621. By this time, most of the historical issues surrounding Proclus, not to mention Euclid of Megara, had been thoroughly resolved. As, one after another, the facts supposedly relevant to the composition of the *Elements* were shown to be false, scholars were left with very little to say about the historical development of the *Elements*, or even about the *absence* of significant development, as Savile wished to do. He thus resorted to a more rhetorical approach, introducing a character who had not yet appeared in any of the discussions of Euclid: Theon's daughter, the mathematician Hypatia. I argue that he employed a metaphor that had been current since the time of Proclus – the

[1] In the very title of his influential 1775 edition of Euclid, Robert Simson claimed to have corrected the errors by which Theon had "long ago vitiated these books," and attributed to him "or whoever was the Editor of the present Greek Text" any infelicity that Simson found in Euclid's reasoning or presentation ((Euclid 1775, p. v), where there is also a brief summary of the controversies over the authorship of the *Elements*). For the modern assumption that any failings in Euclid's work can be attributed to Theon, see especially Wilbur Knorr's critique of Albert Lejeune and other modern scholars of optics, whom he accuses of falling into precisely this trap (Knorr 1994).

*Elements* as a beautiful human body – in order to provoke horror at the sight of Ramus's "dismemberment" of the text, a horror he underlined by reminding his readers of the fate of Hypatia, torn apart by a barbaric mob.

Proclus appears under more than one guise in this story. Renaissance scholars' confusion over his historical identity would be as momentous, in its own way, as their puzzlement over the biography of Euclid himself. But Proclus was also, of course, the author of the single most important source for the history of mathematics available to Renaissance readers. As a misplaced historical actor he was used to separate Euclid and Theon; but (more than a little ironically) in his own writing he praised most eloquently the *unity* of the *Elements*. In his *Commentary*, Proclus was the first to use the metaphor of the *Elements* as a beautiful body; and it is with that text that we shall begin the tangled tale of Euclid and his (more or less) faithful "commentator," Theon.

## Proclus and the Beautiful Body of the *Elements*

Proclus was aware that mathematics had a history. In his *Commentary on the First Book of Euclid's* Elements, he included summaries of the earlier historical writings of Eudemus and others. These provided Renaissance scholars with a chronological narrative of the development of mathematics and remain a crucial source for historians to this day. Proclus also knew that the *Elements* had a history. He was careful to ascribe individual propositions of the *Elements* to particular mathematicians who preceded Euclid: Thales or Pythagoras, for instance.[2] Nevertheless, his position in the *Commentary* was that Euclid had created in the *Elements* a work that transcended its historical roots and was itself above the vicissitudes of history. Euclid achieved this in part through his own peculiar genius, but equally through his place in history, at the culmination of a long Platonic tradition of mathematics. The last in a series of "elementators," Euclid brought their work to a perfection that could not be improved upon – a perfection that befitted Proclus's high Platonic notion of mathematics. Proclus affirmed that Euclid excelled both in choosing the matter to be included in the *Elements*, and in its arrangement. By means of his demonstrations, he had linked the entire structure of his collection into a whole that could not be altered without damaging its almost miraculous perfection:

> We mark also the coherence of its results, the economy and orderliness in its arrangements of primary and corollary propositions, and the cogency with which all the several parts are presented. Indeed, if you add or take away any detail whatever, are you not inadvertently leaving the way of science and being led down the opposite path of error and ignorance?[3]

---

[2] Proclus attributes to Thales the discovery that the diameter bisects the circle (part of the definition of the circle; see Proclus 1992, p. 124). He also assigned to him I.5 (*ibid.*, p. 195); I.15 (p. 233); and I.26 (p. 275). Theorem I.47 he of course attributed to Pythagoras (p. 337).

[3] Proclus (1992, pp. 57–58).

Such was the inviolable unity of the *Elements*, that it deserved to be called an "impeccable and complete exposition" of geometry; its completeness was of a kind that it was beyond criticism, free of any flaw.[4]

The type of perfection exemplified by the *Elements* is what Proclus calls "beauty" elsewhere in the *Commentary*, and the beauty of the *Elements* is a theme that will recur in subsequent debates. Consequently, it is worth taking a moment to consider Proclus's notion of mathematical beauty in some detail. First, however, we need to clarify a problem of interpretation. It is likely that Proclus intended his first prologue (where the discussion of beauty analyzed below is found) to introduce a series of commentaries, which would treat the first several books of Euclid's *Elements* as well as elementary arithmetic. But Proclus completed only the commentary on the first book of the *Elements*. He then, it seems, decided to let the general introduction stand unmodified as the first of two prologues to that commentary.[5] An odd consequence of this editorial decision is that Proclus, in his first prologue to the *Commentary on Euclid's* Elements, does not in fact mention Euclid or the *Elements* at all. Instead, this prologue considers mathematics at the higher level of *mathesis universalis*, the reasoning common to both arithmetic and geometry. But even if the first prologue does not mention the *Elements* explicitly, nevertheless the *Elements* would have formed the principal focus of Proclus's projected mathematical encyclopedia. Proclus certainly had the *Elements* in mind as he wrote on the nature of mathematics in general, and his remarks on mathematical beauty are best understood with respect to the *Elements*.

So, what did Proclus mean by "beauty"? According to his predecessor Plotinus, beauty in all its forms was nothing other than participation in the unifying form of Beauty.[6] But this pure, Platonic notion of beauty would have served Proclus very poorly in defining what made mathematics beautiful. Instead, Proclus proposed a definition of beauty that he attributed to Aristotle: beauty, in body or soul, arises from order, symmetry and definiteness – a notion of beauty from arrangement of parts that Plotinus had explicitly rejected.[7] Proclus argued that bodily ugliness arises from physical disorder and the absence of symmetry and shapeliness – hence the opposite qualities are the components of beauty. Mental ugliness is analogous: an ugly mind is disordered and resistant to the principles of reason. Thus a beautiful mind embraces these very principles: order, symmetry and definiteness.

---

[4] *Ibid.*, p. 58.

[5] See Euclid (1926, vol. 1, p. 32); Proclus (1992, pp. lv–lvi and pp. 344–345).

[6] Plotinus, *Enneads* I.6.

[7] As Glenn Morrow notes, Proclus most probably meant to refer to one of the lost "exoteric" works, such as the *Protrepticus*. See n. 49 on p. 22 of (Proclus 1992), But see also Aristotle, *Metaphysics* 13.3 (1078b1): "The chief forms of beauty are order and symmetry and definiteness, which the mathematical sciences demonstrate in a special degree." Of course, Plotinus did himself stress that what is beautiful about participation in beauty is *unity*. Proclus has expanded this simple unity (via Aristotle) into symmetry, order and definiteness, all of which can be ultimately seen as types of unity in itself (definiteness), with respect to its outgoing expression from its source (order) and in its relation back to its source (symmetry) – the three ontological moments typical of late Neoplatonism.

Such principles, Proclus argued, are found nowhere more completely than in mathematics. Mathematics exemplifies order by explaining more complex propositions by means of simpler ones, ordering its results so that later propositions are always dependent on their predecessors. The symmetry of mathematics is found in the accord of the theorems with one another, and "in their common reference back to Nous" – which is the end to which mathematics directs the minds of its students. Its definiteness, finally, is found in the certainty of its truths.[8] These are all qualities that the *Elements* possess to the highest degree, as Proclus made clear in the passage cited above, where he praised the certainty and orderliness of the *Elements*.

Proclus's description of mathematical symmetry here in the first prologue is a little obscure. What does he mean by saying that theorems "refer back to Nous?" We can gain some insight by considering his actual commentary on the *Elements*. There, he draws the reader's attention not only to the validity of the demonstrations (and hence the truth of the theorems) and their reliance upon each other, but also to their origin in the intelligible realm. For almost all definitions and postulates, Proclus showed that the mathematical objects which we encounter in our imagination or on paper were spatialized representations of the architecture of Being (that is, of Nous or the intelligible realm). Mathematics is not so much about mathematical objects, Proclus thought, as it is a projection into the imagination of deep structures in the highest, most abstract realms of Nous. The study of mathematics is valuable more than anything because it aligns the movements of the highest reaches of our souls with those of Nous.

Proclus's notion of mathematical symmetry as a kind of conformity to Nous is reflected very clearly in his commentary to *Elements* I.1. At the end of his proofs, Proclus observed, Euclid generally restated the problem or theorem that was posed in the enunciation, adding "being what was required to show" and "being what was required to do" (QED and QEF). He did this, explained Proclus, so as to "[join] the end to the beginning in imitation of the Nous that unfolds itself and then returns to its starting point."[9] That is, the movement of the reason as it follows each proof is a kind of imitation of the movement of Nous; a perfect symmetry holds between the mathematical text and the intelligible realm.

In sum, the order, symmetry and definiteness of the *Elements* were what made them – like all mathematics – beautiful. It is important for us to observe the force of the analogy between physical and intelligible beauty. Proclus was answering critics of mathematics who said that the art did not explicitly concern itself with aesthetic

[8] Proclus (1992, pp. 22–23).

[9] Proclus (1992, p. 164). To give another example, at Proclus (1992, p. 88), Proclus glosses the definition of a straight line by explaining that it stands for unwavering providence, while the circle represents activity returning to itself, so that "the demiurgic Nous has ... set up these two principles in himself, the straight and the circular, and produced out of himself two monads, the one acting in a circular fashion to perfect all intelligible essences, the other moving in a straight line to bring all perceptible things to birth." Proclus does not intend these references to Nous to be merely analogies. As he puts it most strikingly at Proclus (1992, p. 113), the geometer's goal is ultimately to leave behind the spatially-extended mathematical objects and return to their models in the intelligible realm, having understood through the relationships of geometrical objects the relationships that obtain among the non-spatial intelligibles.

matters. Proclus's answer was to show (*via* Aristotle) that *physical* beauty was characterized by certain arrangements of features, and that a definition for mental beauty could be found by analogy. Mathematics did not treat beauty as a subject for discourse, but exemplified it through the form it necessarily took. It should be noted, moreover, that Proclus did not discover this form in, say, the act of generation of new mathematical ideas or even in mathematical objects or propositions themselves, but in their presentation as a synthetically reasoned treatise. That is, it is mathematical *texts* that are beautiful, and they are beautiful more because of the perfection of their form, than by virtue of the nature of their contents.

Consider, for example, theorem I.47 of the *Elements* (Pythagoras's theorem). According to Proclus's account, the beauty of this theorem would lie not in the relationship it discloses between the square on the hypotenuse and those on the other two sides of a right-angled triangle. Rather, it would be beautiful because it is definite (from the certainty of its demonstration); ordered (because it is in its right place at the culmination of the first book of the *Elements*, drawing on most of the theorems proven there, and as a necessary precursor for numerous theorems in the subsequent books); and symmetrical (because the relations between lines, triangles and squares revealed in the proof could, in principle, be interpreted as a statement about the structure of the intelligible realm).[10]

In other words, mathematical beauty is found in the text *because* it is a complete, interconnected web of mathematical truths.[11] Moreover – and highly significant for the later debate – propositions and demonstrations have to be considered *together*, not separately. Demonstrations are the source of all three components of beauty: order, symmetry and definiteness. For Proclus, a collection of bare unproven propositions would simply not satisfy the requirements for beauty.

In Proclus's account of beauty, then, the propositions of the *Elements* and their proofs form a single, ordered and beautiful body. No matter that Euclid inherited much of the matter of the *Elements*, from his predecessors. The text that Proclus had in front of him was, in every significant way, the work of the single person Euclid. He had chosen the propositions and put them into miraculous order. Moreover (and most significantly for us), Proclus insisted that Euclid himself had provided the demonstrations for his propositions, and had done so using masterful logic:

> He also included reasonings of all sorts, both proofs founded on causes and proofs based on signs, but all of them impeccable, exact and appropriate to science. Besides these, the book contains all the dialectical methods: the method of division for finding kinds, definitions for making statements of essential properties, demonstrations for preceding from premises to conclusions, and analysis for passing in the reverse direction from conclusions to principles, The various forms of conversion, both the simple and the more complex, can be accurately learned in this treatise.[12]

---

[10] This demonstration of the beauty of I.47 is not found in the *Commentary*. Proclus provides a detailed discussion of the relationship to Nous only for the definitions and postulates. In his commentary on the theorems, he very rarely indicates how precisely the analogy is to be drawn.

[11] For a similar modern approach to mathematical beauty, see Netz (2005, pp. 282–283). Netz argues that the beauty of Pythagoras' Theorem comes not so much from the fact itself, but from the "promise of a narrative" (of proof and supporting propositions) that it holds out.

[12] Proclus (1992, p. 57).

The *Elements*, by virtue of its demonstrations, was a veritable gymnasium for the reasoning faculty; and, as Proclus stated quite unambiguously, these remarkable demonstrations, in all their variety, were the work of Euclid. As author of both the propositions and demonstrations, he had brought definiteness, order and symmetry to the elements of geometry, fashioning an indivisible beautiful body out of the raw material of elementary mathematics.

## The "Commentaries" of Theon

Despite Proclus's insistence that Euclid had written the entire text of the *Elements*, and that the propositions and demonstrations were inseparable from each other, sixteenth-century mathematicians nearly all believed that Euclid had written only the statements of the propositions, and that Theon had written the demonstrations found in the Greek text of the *Elements*. Just as in their confusion over the identity of Euclid, Renaissance scholars seized on ambiguities in the available historical sources, and drew entirely the wrong conclusion about the authorship of the text.

How did Renaissance scholars arrive at this mistaken conclusion? Most Greek manuscripts of the *Elements* attested to Theon's involvement in preparing the text, stating that they were copied *ek tês Theônos ekdoseôs* (from the edition of Theon) or *apo sunousiôn tou Theônos* (from the lectures of Theon). Moreover, in his commentary to the *Almagest*, Theon himself laid claim at least to the demonstration of proposition VI.33 (or the second part of it, which concerns sectors in equal circles): "But that sectors in equal circles are to one another as the angles on which they stand has been proved by me in my edition of the *Elements* at the end of the sixth book."[13]

The regular manuscript references to the "edition" or "lectures" of Theon, together with his own testimony about VI.33, provided the principal (indeed, the only) evidence that Theon had composed the proofs to the *Elements*. These references were hardly sufficient to make the case. Proclus most likely found the same headings in his manuscripts of the *Elements*, yet he never doubted that the work was Euclid's. The medieval transmission of Euclid's text and the new humanist version of Euclid published by Bartolomeo Zamberti in 1505 together explain the peculiar interpretation of this evidence in the sixteenth century.

## Campanus and Zamberti

Until the early sixteenth century, the most common Latin version of the *Elements* in circulation was that of the thirteenth-century mathematician Johannes Campanus of Novara.[14] His was not, however, a literal translation of the text by any means.

[13] Theon (1936, vol. 2, p. 492); Euclid (1926, vol. 1, p. 46).

[14] A complete edition of Campanus's text is now available in Campanus (2005).

Campanus took over most of the enunciations of the propositions from one of the Arabo-Latin translations in circulation. To these he added proofs freely adapted from Euclid's text, or of his own invention. Moreover, in an effort to make his version of the *Elements* a comprehensive mathematical textbook, Campanus added much additional material from such works as Jordanus de Nemore's *Arithmetica*, as well as commentaries on Euclid and other versions of the text – even the original Greek.[15] His demonstrations truly were extended commentaries upon the propositions, and so the manuscripts described them.

The first printed edition of Euclid (Venice, 1482) reinforced the author-commentator relationship between Euclid and Campanus. In the title, the printer Erhard Ratdolt (or his editor) described his edition of the Campanus Euclid as "that most famous work, the *Elements* of Euclid of Megara, together with the commentaries on the geometrical art by the most perceptive Campanus;"[16] and at the very end of the *Elements*, he wrote, "here ends the *Elements* of the geometrical art by Euclid of Megara, and also the commentaries on it [that is, the *Elements*] by the most perceptive Campanus."[17] Ratdolt distinguished quite precisely between the work of Euclid (consisting of the enunciations of the propositions), and that of Campanus (the demonstrations), which he considered to be a commentary on Euclid's *Elements* itself.

When Zamberti published his humanist Euclid, translated directly from the Greek text, he made it clear that he had been inspired to undertake the task because of the flaws in the Campanus version. The title of *his* edition (Venice, 1505) announced that he was providing

> the thirteen books of Euclid's *Elements* with the exposition of the great mathematician Theon. Many things that were missing in the translation of Campanus have been added from the Greek text, and many things that were disordered and absurd have been returned to order and corrected.[18]

From the title alone, it can be seen that Zamberti considered that Theon "expounded" the text of Euclid in just the same way that Campanus had (but to a much higher standard). Yet, just as Zamberti had set side-by-side contradictory reports of the life of Euclid, so too, in his preface to the *Elements*, he put forward two quite different accounts of Euclid's role in composing the *Elements*.

---

[15] Campanus (2005, p. 32).

[16] Euclid (1926, p. 97): "Preclarissimum opus elementorum Euclidis megarensis una cum commentis Campani perspicacissimi in artem geometriam incipit feliciter." See p. 120 above for Ratdolt's edition of Euclid.

[17] *Ibid.*: "Opus elementorum euclidis megarensis in geometriam artem, in id quoque Campani perspicacissimi Commentationes finiunt.

[18] Euclid (1926, p. 98): "Euclidis ... elementorum libros xiij cum expositione Theonis insignis mathematici. quibus multa quae deerant ex lectione graeca sumpta addita sunt nec non plurima perversa et praepostere voluta in Campani interpretatione, ordinata digesta et castigata sunt..." See p. 118 above for this edition and the life of Euclid prefaced to it.

Early in his preface, Zamberti asserted that Euclid collected and marvelously arranged things that had been discovered "by various philosophers."[19] Later on, he reiterated this by quoting from Proclus's brief history of mathematics in his *Commentary*:

> Now Euclid, as I say, was a man of extraordinary ability, who gathered the *Elements* into one. He took much from Eudoxus and Theaetetus, and demonstrated the things he had taken from here and there more easily and clearly than his predecessors had done.[20]

Zamberti's purpose in citing this passage was to show that Euclid *collected* material for the *Elements*. But Proclus made it quite clear – in the very passage Zamberti cited from his *Commentary* – that Euclid also *demonstrated* the theorems he collected. Zamberti even went on to quote at length Proclus's praise of Euclid for incorporating into the *Elements* every kind of scientific reasoning.[21] He was thus fully aware of Proclus's assumption that Euclid had written the whole of the *Elements*, enunciations and demonstrations alike.

It is perplexing, then, that Zamberti presently goes on to state precisely the opposite, in the context of his critique of Campanus. The medieval version had not, he wrote, been so much translated as "crapped out" (*excecata*). Campanus put the *Elements* into disorder, and ruined and corrupted the text so that it was "a chaos, rather than elements."[22] Zamberti's version of the *Elements*, by contrast, was altogether different. He had brought the "complete, pure and perfect" text of Euclid from Greece to Italy, "together with the teaching of Theon."[23]

Zamberti's distinction between the "perfect text" of Euclid and the teachings of Theon was the first time anyone unambiguously attributed part of the *Elements* to Theon.[24] Zamberti joined the consensus that Euclid wrote the propositions in the Campanus version, while Campanus himself wrote the demonstrations – not an entirely inaccurate description, as we have seen. Campanus's edition, however, was quite inadequate, and would be replaced by Zamberti's translation directly from the Greek manuscripts. Zamberti, it seems, then proceeded by way of analogy: if Campanus's version of "Euclid" (who wrote the propositions) was to be replaced by the genuine Euclid, then Campanus the commentator (responsible for the demonstrations) would be replaced by the genuine *Greek* commentator he found in the manuscripts, whom Zamberti identified with Theon, presumably on the basis of the subtitles of the manuscripts which identified the work as "from Theon's lectures."

---

[19] Euclid (1505, fol. 5r): "a diversis philosophis."

[20] For the Latin, see n. 24 of fifth chapter.

[21] Proclus (1992, p. 57). Cited above at n. 12.

[22] Euclid (1505, fol. 5v): "Elementa igitur huiusmodi a Campano non intepretata communi iudicio sed barbarie excecata, praepostere ac perverse subvoluta, corrupte et inscite subversa, et adeo ut non elementa sed accommodatius chaos appellari possint intuentes."

[23] *Ibid.*: "... ut tanta cognitio tandem e graecia Italiam petens integra, pura et perfecta una cum Theonis traditione latinis legenda praeberetur."

[24] De Morgan, in his brief survey of manuscripts and editions of the *Elements*, was also unable to find any reference to Theon's authorship of the demonstrations earlier than Zamberti. See de Morgan (1870, p. 71a).

In other words, despite his earlier citation of Proclan passages in which Euclid appeared as demonstrator as well as collector of propositions, Zamberti's desire to contrast his version of the text with Campanus's commented text led him to split the authorship of the Greek *Elements* in precisely the same way that it was divided in the medieval version.

A little later in the preface, addressing his patron Guidobaldo, duke of Urbino, Zamberti laid out the division of labor more explicitly. If Guidobaldo devoted his attention to this text, Zamberti told him, he would receive gifts greater than all the gold of Araby as he "reads the problems and theorems of the author himself, gathered with miraculous consideration and judgement, and redacted into a unity."[25] Note that Zamberti credited Euclid ("the author himself") only with the gathering of the problems and theorems, and their redaction into a single work. He said nothing at all of the demonstrations, which he went on to assign unambiguously to Theon, whom he praised in terms even more lavish than Euclid:

> And you should see how great was the perception, skill and learning of the commentator, Theon, who explains the sublime sense of the problems and theorems in quite marvellous order, and makes them clear through his investigations. Through the preliminary specification, he sets out what the questions demand. Through the construction, he constructs and builds up marvellously that which is said [in words]. Then, in the proof, he proves the question, laying it out to the senses. And finally, he closes with a conclusion that is both valid and most stable, tying it up so tightly that one would hardly dare to deny what has been proposed and then proven.[26]

In this description of the "commentator's" work, Theon is held to be responsible for the specifications, constructions, proofs and conclusions of the *Elements* – which, taken together, are nothing other than the demonstrations as a whole.[27]

Zamberti went on to extoll the order of the *Elements*, the achievement, he thought, of the author Euclid, who also selected the definitions and postulates. He stressed the organic unity of the *Elements*: "from this point [the beginning of the first book] the teaching of Euclid unfolds itself from the first book to the thirteenth, so that, just as each preceding problem and theorem opens the way for the next problem and theorem, so the first book reveals the second, the second the third, and

[25] *Ibid.*, fol. 6v: "legesque ipsius auctoris problemata et theoremata miro examine et iudicio collecta in unumque redacta."

[26] *Ibid.*: "Videasque quanta sit acuitas, quantum sit ingenium, quantaque doctrina Theonis ipsius interpretis, qui miro quoddam ordine sublimes problematum et theorematum sensus explicat, magnaque indagine patefacit, per prodiorismum nanque ea quae in quaestionibus posita sunt proponit; per constructionem ea quae dicuntur construit et mirabiliter aedificat; inde per demonstrationem comprobat sensui subiiciens, postmodum conclusione firmissima et valida claudit, et astringit adeo ut ea quae proposita et comprobata sunt minime negare audeamus."

[27] According to Proclus (Proclus 1992, pp. 159–160), there are several distinct parts to a theorem or problem, and Zamberti has listed them all here, with the exception of the enunciation (*protasis*) and exposition (*ekthesis*). The "exposition" is so inseparably joined to the specification (*diorismos*) that Zamberti surely meant to refer to both with the term *prodiorismos*. The enunciation, on the other hand, is simply the statement of the proposition; the other parts listed by Zamberti (and attributed to Theon) are what collectively make up the demonstration.

one finds a marvellous order right to the end of the book."[28] Zamberti emphasized his own fidelity to the pure intent of the author – he had neither added nor subtracted any propositions, and had retained the precise order of the Greek – so that the reader could be assured that Zamberti had avoided the errors of Campanus.[29]

Thus Zamberti unambiguously attributed the demonstrations in the *Elements* to Theon, leaving the choice of propositions, their wording and order, and the assembly of definitions and postulates to Euclid. This model of joint composition quite contradicted the passages Zamberti earlier copied out of Proclus. There Euclid not only chose and ordered the results of his predecessors, but also proved them, using and displaying all the tools of dialectic. The example of Campanus and his own rivalry with that edition encouraged him to read back into antiquity the same division between author and commentator that was found in the medieval Euclid; by praising Theon so extravagantly, he set Campanus's contribution to the *Elements* in so much the poorer a light.

Zamberti's division of authorship might seem to challenge Proclus's notion of the unity of the *Elements*. Paradoxically, however, Zamberti was *also* committed to an essentially Proclan view of the "marvellous" nature of the *Elements*: that both through its arrangement *and* its theorems it was perfect and unified to the highest degree. His assertion of dual authorship did not at all imply (for him, at least) that the text was even in principle able to exist in any form other than that in which had actually been transmitted to his age. It was not possible to rearrange the text, or even substitute new demonstrations (as Campanus had done) without serious damage to the "pure and perfect" *Elements*.

## The Divided *Elements*

Nevertheless, by popularizing the notion that Theon had written the demonstrations to the *Elements*, Zamberti opened up the possibility that the text could be divided and altered – sometimes in ways that Zamberti himself, with his belief in the miraculous unity of the *Elements* and admiration for "Theon," would have deplored. Several editions of Euclid were published without the demonstrations, a cost-saving measure that now seemed justified on historical grounds.[30] Even in complete editions of the *Elements*, it became common practice to place Theon's name as a heading above each demonstration, emphasizing his subordinate role as commentator. Such was Johannes Hervagius's 1537 Basel edition of Euclid (itself

[28] Euclid (1505, fol. 6v): "unde omnis Euclidis doctrina a primo volumine usque ad tertium sese extendit decimum, utque sicut theorema et problema praecedens subsequens et theorema et problema aperit, sic primum volumen secundum enodat, et secundum tertium, et sic sequendo usque ad calcem mirabilis ordo invenitur."

[29] *Ibid.*

[30] Among those editions was, of course, Ramus's of 1545 or 1549 discussed in second chapter (see p. 27). On these reduced editions, see (Thomas-Stanford 1926, pp. 2, 11–12).

based upon Jacques Lefèvre D'Étaples's 1516 Euclid, which divided the text up into "commentaries of Theon," "commentaries of Campanus" and "commentaries of Hypsicles").[31] In 1544 Oronce Fine, the regius professor of mathematics at Paris, republished Zamberti's translation and went so far as to replace *all* of the proofs in the *Elements* with his own inventions, heading each demonstration "Orontius" after the example of editions like Hervagius's that labelled the proofs "Theon."[32]

Jacques Peletier, one of the most perceptive readers of the *Elements*, provided a list at the beginning of his 1557 edition which itemized the improvements he had made over previous printings of the text. The list began:

> I have added new demonstrations in every part to Euclid, which I have drawn from the firmest of proofs, the straight line and the equal, and especially from the circle, which is the archetype of the whole of geometry. I have emended some demonstrations of Theon and Campanus, when they did not provide a sufficiently convincing or appropriate proof; others I have made more elegant or more clear.[33]

Peletier's opinion on the state of the text was rather subtle. He had no doubt that the demonstrations in the *Elements* were the work of authors later than Euclid and could, when necessary, be replaced by better demonstrations, so as to establish the truth of Euclid's text more clearly. Nevertheless, he did not believe that Euclid was *simply* a collector and organizer of mathematical truths. Rather, Euclid was a mathematician in his own right, and must have shared the mathematician's concern with proof. Peletier makes this point in a letter addressed to Jean Fernel, published in his edition of the *Elements*, in which he anticipated how critics might view his robust editorial methods. He imagined that some would censure his liberal use of alternative demonstrations, where he drew from other mathematicians as well as his own ingenuity to find the best means of proof. But, he replied,

> those who accuse us in this way should also accuse Theon and Campanus. There is almost nothing from Campanus that was not shared with Theon, and Theon himself put together the proofs of others that fell into his hands. How do I know? Do we really think that Euclid put in order geometrical propositions that he hadn't already confirmed by means of his own proofs? Do we really think that that famous theorem, attributed to Pythagoras of Samos, about squares on the sides of a right-angled triangle, would have come down to us if it weren't backed up by some proof? What do you think Thales before him did? What about those who came after him – Plato, Hippocrates of Chios, Archytas of Tarentum and the whole race of geometers – what do we think *they* did? And finally, do we believe that Euclid would have left out that problem of duplicating the cube, proposed to him by the oracle, if there had been a demonstration?[34]

---

[31] Euclid (1537); Euclid (1516).

[32] Euclid (1544). The "Greek text of Euclid" referred to in the title to this book (*una cum ipsius Euclidis textu graeco*) was only that of the propositions; none of "Theon's" proofs were reproduced.

[33] Euclid (1557, sig. A2r): "Novas Demonstrationes passim ad Euclidem adiecimus: quas ex firmissimis probationum, Recto et Aequali, maxime ex Circulo, totius operis Geometrici archetypo, deprompsimus. Demonstrationes nonnullas Theonis et Campani, quum non satis probabiliter, aut non satis apposite confirmarent, emendavimus; caeteras concinniores clarioresque reddidimus."

[34] *Ibid.*, sig. p4v: "Qui enim nos accusabunt, non iam Campanum, sed Theonem ipsum accusent oportet: quorum ille alter nihil fere habet, quod a Theone non sit mutuatus; hic vero ipse aliorum

So, although Peletier granted that Euclid wrote the enunciations, and Theon and Campanus the proofs, nevertheless he maintained that the latter substituted their proofs for ones that they found in the text – even proofs by Euclid himself. Such originally Euclidean proofs must have existed, because at every stage of its history, mathematics had relied on demonstration. Pythagoras must have had some proof for his great theorem, even if it was not the one we read today in the *Elements*. And Euclid would certainly have included in the *Elements* the method he supposedly discovered for doubling the cube had he also been able to prove that it was mathematically correct.[35] Peletier's own editorial work, then, was only a continuation of a practice that stretched back to the beginning of mathematics. In replacing the proofs of the *Elements* he was not altering Euclid's text, only Theon's additions; and he was doing precisely the same thing that Theon himself had done.

In his 1566 edition of the *Elements*, to give a final example, François de Foix, comte de Candalle assumed that the enunciations of the first thirteen books of the *Elements* had been written by Euclid, with separate sets of commentaries by Campanus and Theon. However, in his edition, he also published the spurious fourteenth and fifteenth books, containing more advanced solid geometry, which the manuscripts attributed to Hypsicles.[36] So convinced was de Foix, however, of the strict separation between author and commentator throughout the *Elements*, that he attributed the *propositions* of the fourteeth and fifteenth books to Euclid, and only the "commentary" (that is, demonstrations) to Hypsicles. Thus, taking into account also the medieval version, there were now *three* commentators on the Euclidean text – a circumstance that worried de Foix:

> But because Theon only transmitted [the proofs] of the first of these three, and Hypsicles the rest, while Campanus wrote on all of them, I fear that these differences may have brought about some corruption.[37]

De Foix was as convinced as Zamberti that the *Elements* was a mathematically unified book – or, at least, it *had* been before the commentators had distorted it. He explained to his readers the marvelous structure of the *Elements*, in which the definitions are simple and clear, the postulates do not demand more than can be readily

---

probationes per manus traditas congessit. Quid enim? an existimamus Propositiones Geometricas ab Euclide in ordinem esse redactas, quae non ante suis assertionibus confirmatae essent? an vero Theorema de laterum Trigoni Rectanguli potentiis tam celebre, a Pythagora Samio relictum fuisse putamus, nisi sua demonstratione munitum? Quid ante eum, Thaletem? quid post eum, Platonem, Hippocratem Chium, Architum Tarentinum, ac totam Geometrarum nationem fecisse putamus? an denique Euclidem Problema illud ab Oraculo propositum de Cubo duplicando, praetermissurum fuisse credemus, si constitisset demonstratio?"

35 An interesting gloss on Valerius Maximus's story of Plato and Euclid of Megara; see text cited at n. 45 of fifth chapter, where Ramus uses this form of argument to cast doubt on the accuracy of Valerius.

36 Hypsicles was in fact the author of the fourteenth book. See Euclid (1926, pp. 14–15).

37 Euclid (1566, sig. ê4r): "Sed quia trium horum priorem tantum transtulit Theon, Ypsicles vero reliquas, Campanus autem in omnes scripsit, veremur has diversitates aliquid corruptionis generasse."

granted, and the common notions are self-evident – a structure still visible despite the depradations of the commentators:

> The propositions, finally, are born from the principles and are arranged on this plan: subsequent ones are to be demonstrated from earlier ones, and not from later ones. The demonstrations, which are to be constructed only by the laws of this art, are embellished with great strictness, lest the use of a mechanical instrument hinder the proof. I see, however, that Euclid's principles and theorems as transmitted to us by Campanus and Theon, violate the strictness of mathematics in their demonstrations.[38]

De Foix went even further and took it upon himself to defend geometry against the ravages of Theon and Campanus:

> In my demonstrations, I shall generally follow the proofs left to us by Theon and Campanus. If, however, they deviate at any time from the purity of geometry, I shall preserve the integrity of this art by using different proofs.[39]

All of these French editors and translators considered the proofs in the *Elements* to be a later accretion, a commentary, added by Theon or Campanus to the list of propositions gathered by Euclid. When these proofs, as they thought, were unworthy of the geometer's propositions, it showed no disrespect to Euclid to alter them; indeed, they were, in a sense, defending him from unworthy commentators. The third editor, de Foix, wrote only a year before Ramus would renew his assault on the *Elements* in his *Prooemium*. De Foix makes a more naive distinction between author and "commentator" than Peletier, returning to the kind of strict division imagined by Fine.

De Foix's encounter with the multiple commentators of the *Elements* (as he supposed them to be) aroused a kind of textual anxiety. On the one hand, the *Elements* for all of these writers was the central mathematical text, the most important product of Greek mathematics. On the other, the received text of the *Elements* was perhaps no longer to be trusted, since it had suffered so many historical vicissitudes. Ramus picked up on this tension over the state of the *Elements* in his *Prooemium* and even more in his *Scholae mathematicae*. He, too, was intensely concerned with the *Elements*; for him, even more than for his predecessors it was the end-product of the entire development of Greek mathematics from Thales to Theon. Yet he turns the suspicion that it may have been botched through its treatment over the centuries into a sustained polemic against the text itself.[40]

---

[38] *Ibid.*, sig. ê3r: "Propositiones demum a principiis genitae, hoc praescripto disponantur, ut subsequentes a prioribus, non autem a posterioribus demonstrandae sint, ac earum demonstrationes solis disciplinae legibus construendae tanta religione decorentur, ne ullum in eis demonstrandis intercedat mecanici instrumenti iuvamen. Quippe Euclidis a Campano et Theone hucusque nobis tradita quaedam principia ac theoremata demonstrationibus, mathematicae religioni repugnare cernimus."

[39] *Ibid.*, sig. ê3v: "Demonstrandi autem argumenta, a Campano et Theone relicta, ut plurimum insequemur, quae si quandoque a Geometrica sinceritate recesserint, in alia demonstrata mutantes, huius disciplinae integritatem tuebimur."

[40] A similar point is made in Loget (2004, pp. 20–22).

## Borrel's Dissent

One of the few voices questioning the attribution of the proofs to Theon was the mathematician Jean Borrel (Johannes Buteo, c. 1492-c. 1564). In an appendix to his 1559 work on the quadrature of the circle.[41] he lambasted both Peletier and Fine for their cavalier approach to the text, in replacing the original proofs with ones of their own devising that frequently did not even make mathematical sense. The very assumption on which they had rejected the original proofs of the *Elements*, was, he said, misguided; it was simply an error, albeit an old and generally-held one, that Theon of Alexandria had written the demonstrations of the *Elements*.[42]

The mistake arose, he wrote, from a misunderstanding of the title found in the Greek codices, *ek tôn Theônos sunousiôn*. a phrase which, he argued, should be translated as *ex omeliis* or *ex expositionibus* ("from the lectures" or "from the expositions"), and not, as these translators had apparently understood it, as *ex demonstrationibus* ("from the demonstrations"); in Greek "demonstrations" would be rendered *apodeixeis*. Borrel argued that it would be unheard of for any ancient geometer to publish his theorems without proofs. Moreover, ancient authors unanimously attributed the *Elements* to Euclid alone, and Proclus in particular testified that Euclid was the author of the theorems and proofs alike. Borrel suggested that the heading referred to a now lost commentary on the *Elements* written by Theon – similar to that written by Proclus; it did not claim that he was responsible for the proofs. Borrel also interpreted Theon's own testimony in his commentary on the *Almagest* in this light. When Theon referred to his *ekdosis* of the *Elements* (a word we should probably translate as "edition"), Borrel said that Theon meant us to understand his commentary on the *Elements*, not the text of the *Elements* itself.

The phrase found at the head of some manuscripts of the *Elements* remained something of a problem. Granted that it could not mean "from the demonstrations of Theon," why should these manuscripts claim to be even "from the commentary of Theon?" Borrel supposed that this had occurred *fraude librarii* – by the deceit of a scribe – who knew the title of Theon's lost commentary to the *Elements* and wanted to pretend that his copy included the commentator's observations.[43]

Borrel made one observation of great significance which was to influence Ramus in his historical criticism of the *Elements*. Many previous authors had cited Proclus as a source for both the philosophy of mathematics and its history. Borrel, for the first time, drew attention to the fact that Proclus could be used as a witness to the text of the *Elements*. By way of explaining just what it was that Theon might have done in his lost commentary, Borrel wrote:

---

[41] Borrel (1559, p. 207, "Annotationum liber in errores Campani, Zamberti, Orontii, Peletarii, Io. Penae interpretum Euclidis.")

[42] *Ibid.*, p. 209: "Vetus est opinio recepta communiter, eas quae Graece leguntur in Elementis demonstrationes non esse Euclidis, sed Theonis Alexandrini."

[43] *Ibid.*, pp. 210–211.

> I would not deny that Theon did some work on the demonstrations in that work ... but I do insist that he did this separately and distinctly, in the course of commenting on some passages. This is just what Proclus did with the first book of the *Elements*. He brought in demonstrations of his own and of others everywhere, but distinguished them from those that we now have in the Greek text by mentioning the author [i.e., Euclid], whom he most often calls the elementator or the geometer, and sometimes calls by his actual name.[44]

Borrel's point was that Proclus quoted demonstrations *as they were now extant in the Greek text*, and attributed them explicitly to Euclid; other material, not found in the received text of the *Elements*, he attributed to other authors or to himself. Borrel made this remark only in passing, but the consequences were clear. Proclus was more than simply a biographer of Euclid. Writing in antiquity and quoting liberally from the *Elements*, he also, quite inadvertently, indicated the state of the text during his lifetime (whenever that was – Borrel does not specify). Borrel's unstated conclusion was that Proclus had read essentially the same text that had reached sixteenth-century Europe, inasmuch as what Proclus attributed to Euclid was what Borrel found in his own text. A few years later, however, Ramus would draw out the consequences of Borrel's remark, expanding it into a tool of criticism: using Proclus as a passive witness to the text rather than an active authority on Euclid. He would, however, come to entirely opposite conclusions to Borrel's.

## Ramus and the Early Date of Proclus

In his *Prooemium* of 1567, Ramus concluded his historical account of the development of the *Elements* of geometry by drawing a dramatic distinction between Euclid and Theon:

> Theon, it seems, far surpassed Euclid, and was the last "elementator." Indeed, the *Elements* of mathematics which are popularly attributed to Euclid should, it seems, be attributed to Theon.[45]

For Ramus (as shown in third chapter) Euclid was just one in a long line of elementators, starting perhaps as early as Pythagoras. By some historical accident, his name alone had come to be associated with the text. In contrast to the historical identity (and authority) that his predecessors had given to Euclid, Ramus intended to reduce the supposed author of the *Elements* to nothing more than an "empty name."[46]

---

[44] *Ibid.*, p. 210: "Non autem negaverim Theonem aliquid demonstrationum in eo opere fecisse ... Hoc tamen dico factum separatim atque distincte inter exponendum locis quibusdam. Quemadmodum et fecit Proclus in primum Elementorum. Nam suas et aliorum demonstrationes passim adducens, an his quas habemus in Graecis libris authoris mensione distinguit quem vel *stoikheiotên*, vel *geômetrên* saepius appellat, interdum etiam nomine proprio."

[45] Ramus (1569, p. 39): "Theon videtur Euclidem longissime superasse, et *stoikheiôtês* ultimus fuisse. Etenim mathematica elementa, quae Euclidi vulgo tribuuntur, videntur Theoni tribuenda."

[46] See text cited at n. 53 in fifth chapter.

Mathematicians from Proclus on had viewed Euclid as a hero of sorts, collecting and putting into perfect order the whole of elementary mathematics. Even if he did not write the proofs, he was responsible for the shape of the *Elements*, its completeness and its structure. Zamberti, we have seen, saw no contradiction between Theon's involvement as a "commentator" and Euclid's being the primary author. The work somehow maintained a unified vision, despite the shared authorship. Moreover, Zamberti was also the principal apologist for the Megaran Euclid; whatever its accidents of composition, the *Elements* commanded authority through its association with Plato.[47] By throwing doubt not only on Euclid's historical identity but also on his very connection with the *Elements*, perhaps even his existence, Ramus undercut any authorial guarantee for the text; it became a free-floating collection of mathematical statements, whose truth was guaranteed only by its supposed logical perfection and rigor. It was precisely Ramus's intention that the text should be so regarded. To complete the case against the *Elements*, he devoted much of the *Scholae* to casting doubt on the legendary Euclidean method, thereby removing (as he thought) any last grounds for retaining it as the central text of mathematical education.

Ramus took the notion of Theonine authorship much further than any of his contemporaries. In Ramus's account, Theon was not simply an editor, or composer of proofs, but an elementator himself – his role was no different from Euclid himself. In order to establish this position, Ramus wished to argue that Theon, as elementator, must have altered the text as a whole – propositions and demonstrations alike – as much as Euclid himself had done; and, seeking evidence for his argument, Ramus believed he could follow Borrel's lead in treating Proclus as a witness to the text of the *Elements*. Immediately after the passage quoted above, he wrote:

> For, among his praises of Euclid, Proclus does not mention the discovery of a single proposition, but only the more careful construction of demonstrations. I have found solid proof of this by comparing the demonstrations in the first book of the *Elements* found in Proclus with any proof by Theon. Proclus, who was earlier than Theon, could neither have seen nor known about the later Theon. Proclus lived in the second century after Christ, Theon in about the fourth. Proclus had Euclid's genuine proofs, in which he sometimes calls Euclid the "Elementator" *par excellence*, sometimes the Geometer, and sometimes by his name, Euclid.[48]

Ramus's debt to Borrel is evident in the last sentence of this excerpt: he quoted him almost *verbatim*, repeating his assertion that Proclus had the true text of Euclid, which Proclus indicated by attaching it to Euclid's name or title. Like Borrel, then,

---

[47] See previous chapter.

[48] *Scholae mathematicae*, p. 39: "Nec enim ullius propositionis inventio inter Euclidis laudes a Proclo numeratur, sed demonstrationum accuratior explicatio. Cuius rei fidem amplissimam nactus sum, comparandis primo Elementorum libro Procli demonstrationibus cum Theonis qualibet demonstratione. Proclus aetate maior Theonem minorem neque videre, neque nosse potuit. Proclus floruit proximo post Christum seculo, Theon fere quarto. Proclus veras Euclidis demonstrationes habuit, in quibus appellatur Euclides per excellentiam modo *stoikheiôtês*, modo *geômetrês*, interdum suo nomine Euclides appellatur."

Ramus called on Proclus as a witness to the text; but Ramus claimed to find substantial *differences* between the text preserved in Proclus's *Commentary* and the received text of the *Elements*. In other words, even if Proclus had the text of Euclid before him, that was not the text that Ramus found in the *Elements* attributed to Euclid. The last person reputed to have altered the text was Theon, who lived (says Ramus) some centuries after Proclus. Since Proclus's genuinely Euclidean text and the version extant in Ramus's time were different, the text must have been changed by Theon. Ramus was thus able to chart an apparent development in the *Elements*, from the second-century recension available to Proclus, to the version bequeathed to posterity by Theon in the fourth century – a development he would flesh out to some extent in the later books of the *Scholae*, corroborating his representation of the *Elements* as a historically shifting text. Yet, as scarcely needs pointing out, Ramus's argument for the development of the *Elements* is based on an evident historical absurdity. For Proclus did *not* live in the second century, some two centuries *before* Theon; he lived in the fifth century, an entire century *later* than Theon.[49]

It would be easy to dismiss Ramus as an incompetent historian and his false dating of Proclus as simply absurd. For one thing, Ramus's chronology renders this late Neoplatonist earlier than even Plotinus, the founder of the Neoplatonic school. But the pseudo-facts about Proclus were widely accepted in the sixteenth century. While Ramus may be guilty of not thinking through their logical consequences, he was not alone in his negligence. The problem of dating Proclus, like that of the identity of Euclid, provides an insight into the means by which historical errors took hold and were propagated in the sixteenth century.

## Ficino on the Platonic School

It might be expected that the revival of Neoplatonism in the fifteenth century would have clarified the history of the late Platonic schools. In fact, Marsilio Ficino, the scholar responsible above all others for the Platonic renewal, had very little to say on the subject. Michael Allen has examined Ficino's remarks on the history of philosophy in a recent book, showing that Ficino approached the problem not as a historian, but as a philosopher, using historical (or quasi-historical) reasoning to very specific philosophical ends. The *origins* of Platonism in the distant past held much more interest for Ficino than did the later history. In his writing on the former problem, Ficino eventually settled on a list of six *prisci theologi*, beginning with Zoroaster and Hermes Trismegistus, and ending with Plato, a pedigree that was meant to assure that Plato's philosophy was not simply the work of a particular man in fourth-century Athens, but a perennial philosophy, dating back to the time of Moses, from whom it was ultimately derived. The names of these ancient theologians were not original with Ficino; other authors (Proclus and Augustine, in particular) had

[49] Ramus's error has also recently been noted by François Loget, in Loget (2004, p. 12). The reasons for the error are much more complicated than Loget suggests, however.

listed some of the members of his ancient succession. But Ficino settled on the complement of six because of the significance of the number: the first perfect number, the days of creation and so forth. While the list was chronologically ordered, Ficino devoted little effort to determining when exactly these semi-mythical figures lived or establishing any historical connections among them, (with the single exception of Plato and Pythagoras, the second to last member of the list).[50] As Allen says, this was "a symbolic, not an historically, let alone chronologically accurate chronology."[51]

Ficino seems to have constructed the history of the Academy to mirror that of the pre-Platonic *prisca sapientia*. In a passage in his *Platonic Theology* that Allen analyzes in detail, Ficino distinguished six Academies, a division made with a clear eye to symmetry. There were three Greek Academies: the "Old Academy," which faithfully preserved Plato's written and unwritten teachings, and two periods of the skeptical "New Academy." The subsequent three Academies not coincidentally fall on the opposite side of the birth of Christ, and represent a return to the positive doctrines of Platonism and a recovery of the lost truths of the Old Academy, revealed and restored through Christian writings such as the Gospel of John and the works of Dionysius the Areopagite. The succession of Neoplatonists, from Plotinus to Proclus and his pupils, took place within these latter Academies (the Egyptian, Roman and Lycian Academies, as Ficino called them). Just as the first Academy in the first group had an authentic grasp on all the Platonic teachings, so too did the first Academy in the second group (the Egyptian Academy of Numenius, Philo and Ammonius Saccas); and just as the second and third Academies of the first group were marred by excessive doubt about Platonic dogma, so were the second and third of the second group misled by being overly attached to the positive doctrines found in the dialogues.[52]

Despite its artificial structure, Ficino's history got the fundamentals correct. In particular, he was very clear on the succession of teachers, from what we would call Middle Platonism through to the closing of the Academy in the sixth century. He also had an accurate grasp of the differences in doctrine among the various Platonic teachers and schools of late antiquity.[53] But in Ficino's strengths lay also his limitations. For his interest was not primarily historical at all, but intellectual and doxographical. He set out the neat structure of Academies in the *Platonic Theology* so as to elucidate a particular Platonic doctrine: the immortality of the soul. He seems to have seen no value in the historical structure itself (and indeed he does not return to this schema of the six Academies anywhere else in his writings). It was not an attempt to understand Platonic authors as actors in a particular historical milieu, any more than was his list of *prisci theologi* a genuine engagement with Egyptian or Greek culture and cultural exchange.

---

[50] Allen (1998, p. 41).

[51] *Ibid.*, p. 25.

[52] *Ibid.*, pp. 70–75.

[53] *Ibid.*, pp. 78–79.

Nevertheless, Ficino was quite certain (and correct) about the order of the later Platonic succession, which he alluded to frequently in other writings. As a careful reader of Proclus, he could hardly have been unaware of the intellectual pedigrees of Platonism; Proclus himself sketched his intellectual ancestry at the beginning of his own *Platonic Theology*,[54] and referred constantly to his predecessors and their intellectual and scholastic connections. But for Ficino it is a *succession*, not a chronology; it conveyed the evolution (in the literal sense of "unfolding") of Plato's doctrine through history, rather than history itself. In a letter he wrote to Cardinal Bessarion, Ficino cited the *Phaedrus*, where Plato said that the gold of wisdom was given by God. Ficino went on to explain that Plato himself was given such divine gold, but wrapped it in obscure words so that it would be valued only by those who understood it. Plato's wisdom remained unappreciated (perhaps an allusion to the skeptical Academy)

> until that gold was brought into the smithy first of Plotinus, then of Porphyry and Iamblichus, and finally of Proclus. There, the dross was removed through an unstinting application of the fire; at last the gold shone forth, filling the entire world with its splendor.[55]

In sum, then, Ficino had a good command of the relative positions of the members of the Platonic school, because, as a Platonist himself, he was interested in the different schools and trends in Platonic philosophy. But he had no interest in their *absolute* historical situation, nor did he ever remark upon it. He knew, of course, that the Platonists in which he was interested lived after Christ; and he commented on that precisely because he wished to detect a Christian influence on their work. But never does he try to establish, for instance, the year of birth of any of the philosophers. If any sixteenth-century author had turned to Ficino to discover what century Proclus lived in, he would have been disappointed. Even his limited historical interest was restricted to simplified lines of teachers and students, through which the Platonic wisdom emerged ever more clearly.

Ficino's version of Porphyry's *Life of Proclus* is quite revealing of where his interests lay. In his translation of the *Enneads*, Ficino translated Porphyry's biography (which was prefaced to the Plotinus's works in the manuscript tradition) without any comment whatsoever. Indeed, although Porphyry had much to say about Plotinus's character, his debates with others and the development of his own philosophical vision – his work, that is, as a philosopher, Ficino completely ignored this

---

[54] Translated in Dillon (2004, pp. 281–282). In this passage (which no doubt influenced Ficino) Proclus also believed there had been a veiling of the Platonic truth at the close of the Old Academy, but it was not rectified until Plotinus, rather than the Middle Platonists, as Ficino would argue.

[55] Ficino (1497, fol. 7r–v): "Verum in Plotini primum, Porphyrii deinde et Iamblichi ac denique Proculi officinam aurum illud iniectum, exquisitissimo ignis examine excussis arenis enituit usque adeo, ut omnem orbem miro splendore repleverit." Note that, in contrast with the model of the six Academies, Ficino offered no criticism here of the late Platonic writers; in fact, it seems that the hidden truth emerges more clearly in the later writers. The sequence of Platonic writers was repeated in another letter, in which Ficino began with the *prisca philosophia* and represented the Platonic school as its witness and consummation. Again, we have the names of the principal late Platonic authors, to which he added the Christians Dionysius, Augustine and Hilary. See Ficino, (1497, fols 104r–v).

in the "Exhortation" to his readers that he placed between the life of Plotinus and the beginning of the *Enneads*. He wrote:

> First, I tell all of you who approach here to hear the divine Plotinus, that you should be aware that you are going to hear Plato himself speaking in the character of Plotinus. Whether it is that Plato has been reborn in Plotinus (which Pythagoreans will surely grant us) or whether the same *daemon* first inspired Plato, and then Plotinus, which no Platonist will deny: absolutely the same inspiring spirit blew into the mouth of Plato as into that of Plotinus.[56]

Ficino saw philosophers in the Platonic tradition to be merely interpreters of the Platonic wisdom. They could be better or worse interpreters, more or less faithful to the wisdom hidden in the dialogues, but they were not creative thinkers in their own rights, still less products of their time. Nevertheless, the chain of teachers and pupils represented a continuous unfolding of ever more refined and truthful interpretation. For that reason, at least, it was essential to establish their order correctly, something which was never in doubt from Ficino on.

## Proclus Out of Time

When humanist historians turned to the chronology of the Platonist school, they thus found little in Ficino to help them beyond the bare succession of philosophers. The first attempt to fill this list out with real historical detail made many significant errors, turning the history of late Platonism into a confusing, contradictory maze.

In 1506, the Volterran humanist Raffaele Maffei published his encyclopedia, the *Commentaria urbana*, including an enormous collection of biographies of philosophers and writers, arranged alphabetically, detailing the extant and lost writings of almost every known ancient author. This list was habitually consulted (though seldom acknowledged) by Renaissance authors and editors wishing to add to a little historical color to their writings, and even by those who had a more serious historical purpose. Yet the work itself was frequently unreliable; in the *Commentaria* we find the seeds of almost all the subsequent confusion over Proclus.

For the entries that are relevant to our problem, Maffei relied largely on the Byzantine encyclopedia known as the *Suda*; from his own wider reading and by inference, he made connections between authors and added dates (which *Suda* itself rarely did). The limitations of Maffei's scholarship became apparent in his attempts to reconcile his disparate sources. Often he drew false connections on the basis of similarity of names alone. In his entry on Proclus, for example, he began well enough by translating the biographical data he found in the *Suda*, which relates that

[56] Plotinus (1492, sig. b2r): "Principio vos omnes admoneo, qui divinum auditum Plotinum huc acceditis, ut Platonem ipsum sub Plotini persona loquentem vos audituros existimetis. Sive enim Plato quondam in Plotino revixit, quod facile nobis Pythagorici dabunt, sive Demon idem Platonem quidem prius afflavit, deinde vero Plotinum, quod Platonici nulli negabunt. Omnino aspirator idem os Platonicum afflat atque Plotinicum."

Proclus was a Platonic philosopher (the head, in fact, of the Athenian Academy), a student of Syrianus, the teacher of Marinus of Neapolis and author of commentaries on Homer and on Plato's *Republic*. Maffei added that he was also the author of several other extant commentaries on Platonic dialogues (but neither he nor the *Suda* mentions the commentary on Euclid). The *Suda* and Maffei both went on to record that Proclus was the second Platonic author, after Porphyry, to write against the Christians, and that he was attacked for this by John Philoponus. Maffei then added a further note that would create no end of subsequent trouble:

> He was also the tutor of M. Antoninus [that is, Marcus Aurelius], who raised him to the consulship, as Spartianus tells us.[57]

Maffei's reference was to the *Historia Augusta*, the notoriously unreliable collection of lives of the later emperors, purported to have been written by Aelius Spartianus and several other authors. The passage Maffei is thinking of, however, comes from the life of Marcus Aurelius attributed to Julius Capitolinus, not Spartianus, where we read:

> Besides these, his teachers in grammar were the Greek Alexander of Cotiaeum, and the Latins Trosius Aper, Pollio, and Eutychius Proculus of Sicca ... he advanced Proculus ... to a proconsulship, though assumed the [financial] burdens of the office himself.

The problem here is not Maffei's failure to recall which historian wrote the biography of Marcus Aurelius, nor even the dubious source from which he obtained his information. Rather, it is his zeal to identify historical characters based on nothing more than similarity of name. We are told here of a *Latin* speaking philosopher called Eutychius *Proculus*, from the north African city of Sicca. Already the imperial tutor has two strikes against him – his nationality and his name – yet Maffei identified him with the Greek Neoplatonist from Asia Minor.[58] On the slenderest of evidence, Proclus, the Platonic philosopher of the fifth century A.D., became tutor to the philosopher-emperor Marcus Aurelius (r. 161–180).[59]

---

[57] Maffei (1506, fol. 259r) The complete entry reads (with the quoted sentence at the very end): "Proclus Lycius discipulus Syriani, philosophus Platonicus, praefuit scholae Atheniensi, cuius discipulus et successor Marinus Neapolitanus fuit. Scripsit plura in philosophia et grammatica. Commentarios in totum Homerum, in Hesiodi Erga et Hemeras, in Rempublicam Platonis; preterea contra christianos Epicheremata XVIII. Hic est Proclus qui post Porphyrium secundus contra nos latravit. Adversus quem Ioannes cognomento grammaticus scripsit apologiam, dictitans eum quanquam in rebus Graecanicis magnum, stultum tamen esse et in hac parte indoctum. Autor Suidas. Is est cuius hodie commentarios in Platonem habemus. Praeceptor etiam M. Antonini, quem ad consulatum usque provexit, ut autor Spartianus."

[58] In defense of Maffei, it should be noted that Ficino often referred to Proclus as Proculus. See quote at n. 55 above.

[59] It may strike us as peculiar that the Stoic emperor is provided with a Platonist tutor. This would not have seemed at all incongruous in this period, however, since Marcus Aurelius was not thought of as a Stoic until Isaac Casaubon's 1605 edition of Persius. Wilhelm Xylander, the editor of the *editio princeps* of the *Meditations* (Xylander, 1559) says nothing about Stoicism in his prefaces or notes to the text, emphasizing only the compatibility of the emperor's thoughts with Christianity. See Kraye (2000). It is worth noting that Xylander included as an appendix to the *Meditations*

If we now cross-reference the other figures mentioned in Maffei's biography, we find the confusions multiplying. Looking up Proclus's student Marinus, we read that he was a philosopher and orator, well-versed in Greek learning, who wrote a life of Proclus "in verse and prose." Thus far Maffei was closely following the *Suda*, from which he also learned that Marinus succeeded Proclus as head of the Academy. Then, departing from his source, Maffei added that Marinus's succession took place *"under the Emperor Hadrian."*[60] The source of Maffei's confusion here was the existence of another Marinus, a geographer whom Ptolemy cites in his *Geography*, who did indeed flourish under the early Antonines.[61] Of course, Marinus the pupil of Proclus could not have flourished under Hadrian, for the simple fact that he lived 300 years later. But because Maffei labored under the misconception that Proclus lived in the second century, an Antonine Marinus may have seemed quite plausible to him. Even so, this statement hardly squares with the other evidence Maffei has just presented. If Marinus succeeded Proclus during the reign of Hadrian (i.e., 117–138), then Proclus would have relinquished his position in the Academy at least 23 years before he became imperial tutor in philosophy. This is not a chronologically *impossible* scenario, but it is highly unlikely.

The case of Syrianus, whom the *Suda* identifies as the teacher of Proclus, is even more complicated. Maffei does not have an entry for Syrianus himself, but he does have one for Syrianus's teacher, Plutarch. Here he differentiated between the Plutarch who taught Syrianus and the Plutarch who wrote the Parallel Lives – for once correctly distinguishing between two philosophers with identical names. The philosopher and head of the Academy flourished, he said, under Julian the Apostate (in other words, in the late fourth century), while the biographer and essayist lived two centuries earlier, during the reign of Trajan. If Maffei had joined the dots here, he would have ended up with an approximately correct date for Proclus himself: if Plutarch held the headship of the Academy in the late fourth century, then his pupil and successor Syrianus must have flourished in the late fourth or early fifth century, placing *his* pupil Proclus, in turn, squarely in the fifth century. Such a conclusion would clearly contradict the date he proposed in his life of Proclus, namely, that the philosopher flourished under Marcus Aurelius. But Maffei left the contradiction unremarked.

Confusion reigns once again when we look at the entry for Hermes. As well as the expected Hermes Trismegistus, we are told of *another* Hermes (mentioned by

---

the text of the life of Proclus "because in its subject matter it is clearly germane to the Emperor's work." (Xylander (1559, p. 2) of the separately paginated Marinus: "argumento ab argumento ab Antonini libris minime alienus.") Xylander saw no incompatibility between the *Meditations* and Platonism; moreover, his association of Proclus with Marcus Aurelius may have added fuel to the early date of Proclus.

[60] [maffei commentaria 1506], s.v. "Marinus." The entire entry reads: "Marinus Neapolitanus philosophus et orator, Graece eruditus, Procli discipulus et successor, sub Hadriano; scripsit ipsius Procli et vitam et dissertationes versibus, ac soluta oratione et nonullas item physicas quaestiones, autor Suidas."

[61] Pauly-Wissowa, *RE*, s.v. "Marinus."

the *Suda*), this one a pupil of Syrianus and thus a "condiscipulus" of Proclus "under Hadrian," according to Maffei.[62] Although he earlier stated that Syrianus lived in the late fourth century, here he puts him back into the second century. Perhaps, while writing the entry for the "other Hermes," Maffei realized that if Syrianus was his teacher, he must have been a fellow student with Proclus. Using what he considered to be his solid dating for Marinus, he made Hermes, too, a contemporary of Trajan, and so, *a fortiori*, Syrianus had to have lived yet earlier.

Maffei's depiction of the Neoplatonist school was hopelessly muddled. He inherited from Ficino the correct line of succession of the late Academy (Plutarch of Athens, Syrianus, Proclus and Marinus), and he often invoked this lineage in order to locate philosophers and their students relative to one another. In locating members of the succession absolutely, however, he vacillated between the second century and the fourth to fifth, but had a marked preference for the earlier, incorrect date. Checking and cross-checking a few entries in the *Commentaria* would give most readers the impression that Proclus lived in the second century.

Several later scholars drew precisely this conclusion. The German mathematician Johannes Stöffler published, in 1534, an edition of the Latin version of the *Sphere*, an elementary astronomical work falsely attributed to Proclus. In his preface, he set out a brief biography of the putative author, writing:

> Proclus of the Lycian nation, pupil of Syrianus, was a Platonic philosopher, the head of the school at Athens. Syrianus of Alexandria, a Platonic philosopher, taught at Athens. His pupil and successor was our Proclus. Marinus of Naples, a philosopher and orator, and well-versed in Greek learning, was Proclus's pupil and successor. He wrote, in the time of Hadrian, about Proclus' life and works, in verse and prose. Therefore, by conjecture, Proclus lived in the time of Trajan or thereabouts.[63]

Stöffler's indebtedness to Maffei is obvious. But perhaps he can be forgiven for not recognizing the errors in his source; Stöffler was no accomplished Greek scholar, as the rest of his preface demonstrates. He puzzled over Proclus's *cognomen* "Diadochus" (which means "successor" to the head of the Academy). Stöffler was convinced, however, that it was a reference to a variety of beryl mentioned by Pliny called *diadochos*, and compared Proclus's surname to that of many German nobles who are named after hard stones or mountains, such as Herttenfelser, Trakkenfelser and Gryffenfelser.[64]

But it was not only second-rate Hellenists like Stöffler who relied on Maffei. Francesco Barozzi, in his 1560 Latin translation of Proclus's *Commentary on*

---

[62] *Ibid.*, fol. 214r: "Hermes alter item, philosophus Aegyptius, auditor Syriani sophistae, condiscipulus Procli sub Hadriano principe."

[63] Stöffler (1534, fol. 1r): "Quartus et est noster Proclus, natione Lycius, discipulus Syriani; philosophus Platonicus, praefuit scholae Atheniensi. Syrianus Alexandrinus philosophus Platonicus, docuit Athenis. Huius discipulus et successor fuit noster Proclus. Marinus Neapolitanus philosophus et orator, graece eruditus, Procli discipulus et successor, sub Adriano scripsit ipsius Procli et vitam et dissertationes, versibus et soluta ratione. Quare iuxta coniecturam, Proclus floruit sub Traiani temporibus, aut circiter."

[64] *Ibid.*, fols 1r–v.

*Euclid*, committed precisely the same error over the date of Proclus, clearly relying on Maffei as his source. It should be emphasized that Barozzi's translation was a masterpiece of scholarship. Rejecting the very poor *editio princeps* of the Greek text, he established his own text from several manuscripts. Although he did not publish his text, modern editors have used the Latin translation to infer Barozzi's excellent readings of the Greek.[65] Yet his scholarship seemed to desert him in his preface, where he wrote:

> I would first like you to know that, although there were several Procluses, one was the most famous, who had the *cognomen* "Diadochus," that is, successor. He was from Lycia, a Platonic philosopher and extraordinary mathematician, who (if we are to believe the *Suda*) was a student of the great Syrianus. When he became head of the Athenian school, he had many students. A notable one was Marinus of Neapolis, and another was M. Antoninus, by whom he was raised to the consulate, as Spartianus records.[66]

Barozzi lifted all of this directly from Maffei's entry on Proclus, only adding a few flourishes here and there to make it look like he had conducted original research, hunting through the *Suda* and the *Historia Augusta* for information on his author. The last sentence, with its confident reference to the incorrect author of the life of Marcus Aurelius, Aelius Spartianus, confirms his reliance on Maffei. It does seem astonishing that the learned Barozzi, having taken the trouble to establish and translate the difficult Greek text of the *Commentary*, should make such a half-hearted attempt at biography. Chronology probably interested him as little as it did Ficino. Nevertheless, in sketching this brief and borrowed life of Proclus, Barozzi gave the early dating scholarly respectability – enough, at least, to smuggle it past the scrutiny of many subsequent, and otherwise critical readers.

## Ramus's Historical Critique of the *Elements*

Ramus almost certainly consulted Barozzi's translation of Proclus's *Commentary* in his work on Euclid's text, and this was the most likely source of his own misdating of Proclus to the second century. His casual identification of Proclus as a second-century philosopher now makes sense. This was no idiosyncratic error on Ramus's part, but a well-established tradition by the time he wrote his *Prooemium* in the 1560s. But Ramus did more than just echo the erroneous biography that others had established. Borrel, in his defense of the unity of the *Elements*, had suggested in

---

65 Proclus (1992, p. lxviii). But see the previous chapter of this book for Barozzi's error over the identity of Euclid.

66 Proclus (1560, sig. **2r): "Primum itaque te scire velim praeter alios multos Proclos, unum Clarissimum omnium fuisse, cognomine Diadochum, hoc est successorem, patria Lycium, Platonicum philosophum, Mathematicumque praestantissimum, qui (si Suidae credendum est) magni Syriani fuit discipulus, cumque Atheniensi Scholae praefuisset, alios ipse discipulos habuit, e quorum numero unus, insignisque fuit Marinus Neapolitanus eius successor; alter M. Antoninus, a quo etiam (ut refert Spartianus) ad consulatum usque provectus fuit."

passing Proclus's value as a witness to Euclid's text, rather than just an authoritative commentator upon it. Ramus took Borrel's suggestion seriously (though Borrel would not have appreciated the conclusions he reached from it). He took what was hitherto no more than an interesting "fact" – that Proclus lived in the second century – and put it to work in the criticism of the *Elements*. That his critique was wrong, built as it was on a false principle, should not be allowed to obscure the novelty and fruitfulness of his approach to the historiography of mathematics. It may have been incorrect, but it was certainly not misguided to look for an early witness to the *Elements*, and to attempt a historical criticism of the text on that basis.

Ramus started from the widespread assumption among his contemporaries that Euclid wrote the theorems in the *Elements*, and Theon the proofs; and in his *Prooemium* Ramus attempted to put these speculations to the test. Since Proclus preceded Theon, he must have had access to a version of the *Elements* untouched by Theon – and this must have been the version he quoted from extensively in his *Commentary*. Thus, according to Ramus, one might judge the extent of Theon's editorial or authorial contribution by comparing his version of the text with the "earlier" witness of Proclus.

Ramus claimed to find that there were many small differences between Proclus's and "Theon's" *Elements*, even if the list of propositions was all but identical in the two versions. Ramus catalogued these in his book-by-book critique of the *Elements*, noting for the most part small changes in wording or order of propositions; but sometimes he discovered an entirely different proof in "Theon" (that is, the text of the *Elements* known in Ramus's day) and "Euclid" (the text quoted by Proclus). For the most part, Ramus was actually documenting changes that had crept into the Greek text through the errors of copyists and the work of Byzantine commentators and scholiasts on the *Elements* as they freely paraphrased and "improved" the proofs of the *Elements*. Sometimes he was observing nothing more than the imperfections of Simon Grynaeus's 1533 *editio princeps*. But for Ramus these small (and, from our point of view, largely illusory) differences led to a series of surprising conclusions. First, that the dominant view about the authorship of the *Elements* was incorrect. If Euclid had written the propositions and Theon the demonstrations, then Proclus would have had before him an *Elements* completely devoid of proofs, or with proofs always different from those found in the modern text. But this, of course, was not the case. Proclus not only cited proofs throughout his *Commentary*, he cited ones that were more or less like those in the modern *Elements*.

Nor could it be argued that Euclid had written the entire text, while Theon made some cosmetic changes (the position argued by Borrel). Ramus had devoted a large part of the *Prooemium* listing elementators, and wringing out any scrap of information he could discover about their activities. Each, he argued, had improved incrementally on his predecessors. There was no precise information about what Euclid had done, apart from Proclus's testimony that he had taken the discoveries of his predecessors and improved the demonstrations and order of theorems.[67] Yet

[67] See the passage quoted at n. 20 above.

this would be a very accurate description also of what *Theon* did to the *Elements*, as Ramus had discovered from his comparison between the versions of the texts. Thus it seemed that Euclid and Theon had each engaged in precisely the same type of scholarly work as all the earlier elementators: editing, rearranging and occasionally augmenting a far older body of mathematical knowledge that had been accumulating since the time of Thales and Pythagoras. In other words, there was nothing particularly special about Euclid; Ramus's dismissal of the prince of geometers as nothing but an "empty name" was apparently justified. Theon was the author of the *Elements*, or rather, he had at least as much claim to authorship as Euclid did to this chaotic, historically accreted text. For this reason, Ramus put Theon at the very end of his account of ancient mathematics in the first book of the *Scholae mathematicae*: both because he knew of no ancient mathematical writer who came later than him (having made Proclus so much earlier a writer), and because the final touches he put on the *Elements* marked a culmination – or the very lowest point – of the Greek abandonment of natural reason.

## Rending the *Elements*

Relying on his discovery of the historical composition of the *Elements*, Ramus separated the *Elements* into its distinct historical members (as he imagined them). He pursued this task more with the gusto of a butcher than the finesse of a surgeon:

> So let us enter right into Euclid's *Elements*, and let us penetrate right into its guts. Let us pull apart the bones, flesh, spirit and blood. Let us discover its hidden causes, so that we can cure the disease we have found.[68]

In other words, the *Elements* would have to be dismembered before it could be repaired to his satisfaction, destroyed in order to be saved. Ramus took up this metaphor again towards the end of third book, where he considered the reasons for the obscurity of mathematics.[69] At times Ramus, ever the versatile polemicist, seemed to blame the mixed-up body of the *Elements* on Euclid (rather than Theon), mocking him for his inept command of logic – the same Euclid he elsewhere dismissed as a non-entity. In the latest-written part of the *Scholae*, however, he more often returned to his historical explanation for its disorder. Going through his proposition by proposition critique of the *Elements* (in the sixth through thirty-first books of the *Scholae*), Ramus took for granted his historical analysis of the text, and his identification of the Proclan text as a more primitive version.

In places, the version corrected by Theon (that is, the common Greek text) was, in Ramus's opinion, better than the genuine Euclid (that is, Proclus's text);

[68] *Scholae mathematicae*, p. 91: "Ingrediamur igitur in ipsa Euclidis elementa, inque viscera ipsa penitus subeamus: sanguinem, spiritum, carnem, ossa retexamus: intimas propositi ad curandum morbi caussas perscrutemur."

[69] Note, for instance, the extended use of the metaphor of members and body at Ramus (1569, p. 102), as Ramus considered the "hysterologia" of the Elements.

in other places it was definitely worse. Theon made one major change to the *Elements* which, from Ramus's point of view, was certainly a mistake. The text Proclus read, claims Ramus, made extensive use of analysis, the logical order that came closest to Ramus's own single, natural method.[70] But Theon destroyed the last vestiges of the natural mathematics that the *Elements* had been built on by removing analysis from all but five propositions. Ramus's claim was based upon Proclus's statement that Euclid included "demonstrations for preceding from premises to conclusions, and analysis for passing in the reverse direction from conclusions to principles" – the often-quoted passage on Euclid's logical perfection.[71] In fact, when Proclus said that the *Elements* contained examples of analysis, he was most likely referring to the very same five propositions as Ramus. In all the manuscripts of the *Elements*, propositions XIII.1–5 contain an appendix explaining the meanings of analysis and synthesis, and providing an analysis for each of these propositions. These remarks on analysis probably date from the pre-Euclidean Academy.[72] Ramus had, in other words, misread Proclus to be saying that analysis was a *common* feature of the *Elements*, and then explained its absence by invoking Theon's heavy-handed editing.

For the most part, though, Theon had a much lighter touch, and his changes (as Ramus saw them) were neutral, or only slightly better or worse than the Euclidean text. An example of an improvement was in his treatment of the fourth and fifth postulates.[73] Ramus agreed with Proclus that the first three postulates certainly were postulates in nature as well as name, but the fourth and fifth were not. Proclus provided several (flawed) proofs of each in his *Commentary*, showing to his and Ramus's satisfaction that they were not self-evident first principles. Proclus's arguments, claimed Ramus, persuaded Theon to move these postulates among the axioms. In fact, however, there was considerable variation in the lists of axioms and postulates in the manuscripts and in the printed traditions, right up to the modern period. Proclus's division between postulates and common notions was certainly the correct one.[74] Elsewhere, Ramus was misled by Proclus's tendency to paraphrase. In his commentary to I.22, Proclus claimed to be quoting Euclid's exact words, but in fact gave an imprecise recollection of the text.[75] Ramus seized upon this discrepancy to corroborate his claim that there are very many small differences between Proclus's text and Theon's:

---

[70] Ramus (1569, p. 99).

[71] See at n. 12 above.

[72] Euclid (1926, vol. 3, p. 442).

[73] Ramus (1569, p. 161).

[74] Euclid (1926, vol. 1, pp. 221–224). Grynaeus's *editio princeps* had the fourth and fifth postulates attached to the end of the common notions; this was no doubt what Ramus was thinking of. The same ordering is found in Simson's influential eighteenth-century Euclid and many of the nineteenth-century English editions that were built upon it.

[75] Euclid (1926, vol. 1, p. 35).

> Proclus cites Euclid *ad verbum* in this demonstration. But his words do not agree with Theon's at all. From this passage and many others it becomes very evident that the text of Theon is not the text of Euclid, but their arguments are generally the same.[76]

Ramus discerned many other such small changes that Theon had made to clarify the language and order of the definitions and principles.[77] A corollary to I.15 that was in Proclus's text, but not in Ramus's Greek text, provided a rare example of Theon's rather more aggressive intervention.[78] Inevitably, Ramus's observations on the differences between Euclid and Theon trailed off after the first book of the *Elements*, since beyond that point he no longer had Proclus's text to rely upon. Yet he continued to try to explain the text historically, using other comparative resources. For example, in his commentary on VI.33, he argued that Theon's proof was deficient, and turned to Euclid's *Catoptrics* for the foundation of a better and, he thinks, older proof.[79] In his commentary on X.5, he quibbled over the terminology that "Theon" used for rational and irrational numbers. To settle the original Euclidean usage he consulted Marinus's commentary on the *Data* – because Marinus, as a student of Proclus's and "thus" a precursor of Theon, also must have had access to the older Euclidean text.[80]

Ramus's historical criticism showed that Theon was not a particularly drastic editor. His only major change was to remove analyses from the *Elements*. Apart from that, he made countless small verbal changes, sometimes for the better, sometimes not, and he rearranged or touched up some of the definitions and postulates. And this, of course, was Ramus's point. Theon neither overhauled the whole of Euclid's text, nor did he compose the demonstrations. Rather, he made many little, unsystematic incremental changes, just as all the previous elementators had done, including Euclid.

Ramus had stumbled upon a text-critical method that might have been very fruitful, had he been more interested in understanding the *Elements* for its own sake. But his approach was always polemical, his intent always to bury the *Elements* (or, at least, alter it beyond recognition), not to elucidate it. His constant references to the small differences between the versions of the *Elements* were meant to remind the reader of its haphazard, historical character and thereby undermine confidence in it altogether. Alongside his hostile intent, there were also some missed opportunities to take a historical approach to the *Elements*. Ramus was so fixed upon his discovery of the two versions of the *Elements* that he gave almost no thought to other ways of

---

76 Ramus (1569, p. 181): "Proclus in hac demonstratione citat Euclidem ad verbum. At verba illa nequaquam cum Theonis verbis conveniunt, ut ex hoc loco, et plerisque aliis notissimum sit, Theonis orationem Euclidis orationem non esse, argumenta tamen plerumque eadem sunt."

77 For example, at p. 155 of the *Scholae mathematicae*, Ramus documents different versions of definition 15, of a circle; and on the following page he tries to establish the movements between the first book and the third of the definition of a segment of a circle.

78 Ramus (1569, p. 178).

79 *Ibid.*, p. 242.

80 *Ibid.*, p. 260.

establishing the text – even though he himself had opened up another very important route to the early text of the *Elements*.

When Ramus redated Euclid to his correct historical position, more than a century later than had been hitherto thought, Aristotle might have become a witness to pre-Euclidean geometry. But despite Ramus's familiarity with Aristotle's logic – he revised and reissued his enormous attack upon it almost every year – he made almost no use of the philosopher's frequent geometrical examples. To give the best-known example, Aristotle's proof of what is now proposition I.5 (the equality of base angles of an isosceles triangle) is entirely different from Euclid's. As Benno Artmann has argued, Aristotle probably had before him Leon's *Elements*, the textbook used in the Academy. The differences between the demonstration known to Aristotle and that found in Euclid's *Elements* reveal Euclid's deep rethinking of the fundamentals of geometry, in particular his banishment of curvilinear angles from the geometer's toolbox.[81] Ramus, in his commentary on this passage, strayed little beyond Proclus's historical remarks (that the theorem was discovered by Thales), devoting most of his effort to quite tendentious logical criticisms.[82]

## Henry Savile on Theon and Proclus

In his life of Euclid in his 1570 lectures, Henry Savile repeated without acknowledgement Ramus's erroneous chronology: "Proclus flourished in the second century after Christ, Theon in about the fourth."[83] Thus, he said, we could read a more primitive version of the *Elements* in Proclus's text. But Savile the Platonist drew from this account of Proclus's life a conclusion quite opposite to Ramus's. Savile found little or no difference between the two versions, at least as far as the first book was concerned (which is as far as Proclus's commentary extended). But, continued Savile, the first book was the most in need of vigorous editing, as was clear from Proclus's frequent addition of additional "cases" to the demonstrations of the *Elements*, and his criticism of the parallel postulate and other fundamental aspects of the book. Relying on his inverted chronology in which Theon preceded Proclus, Savile was certain that Theon, being a responsible editor, must have read Proclus's commentary before starting out on the task of redacting the *Elements*. If he could let the whole first book pass without making any of the obvious changes, then he could hardly have done much to the rest of the *Elements*. Thus, concluded Savile, the *Elements* now extant was much the same as the one written by Euclid; Theon's contribution was trivial.

[81] Artmann (1999, pp. 24–26).

[82] Ramus (1569, pp. 171–172).

[83] MS Savile 29, fol. 42r: "Proclus floruit proximo post Christum seculo, Theon fere quarto."

Savile's acceptance of Ramus's false chronology for Proclus had the effect of telescoping the whole late-antique Platonist school into the first two centuries of the Christian era. As a final argument against Ramus, he wrote:

> Proclus, Pappus, Hero, Simplicius, Philoponus, Alexander, Ammonius, who all preceded Theon, explicitly name Euclid when quoting many propositions which they give the same number as they now bear; they all testify to this same [opinion] on Euclid, author of the Elements, and testify with such necessary force that only a blind man, utterly ignorant of antiquity, would, in the clear light of the evidence, admit even the slightest doubt about the author.[84]

This is an astonishing conclusion, but there is evidence that Savile was very proud of his chronological arguments and took them quite seriously. His student John Chamber owned a printed copy of the *Elements*, into which he copied observations on Euclid from Savile's notes and conversations (all of which he marked with a prominent "S" in a circle). On the titlepage Chamber wrote:

> The proofs are habitually and falsely asserted to be Theon's. It can be conclusively shown that they are Euclid's by countless arguments. In particular, [the text of] Proclus, who lived before Theon, so completely accords with these proofs that it is obvious that Euclid not only furnished his theorems with proofs, but also with *these very* proofs. Let them consult Proclus on proposition 18, at the passage "then the geometer in the..." etc.[85]

Savile returned to the subject some 50 years later, in the lectures on Euclid he gave to inaugurate the Savilian professorship of geometry. In the intervening period, the correct chronology of the late Platonic school had been established. Jean-Jacques Frisius, in his *Bibliotheca philosophorum classicorum* of 1592, provided without any fanfare or explanation an entirely accurate chronology of Proclus and other members of the Platonic school.[86] After Frisius, the false dating of Proclus simply disappeared from scholarly discourse – and so, naturally, did the force of any argument about the state of Euclid's text that drew on Proclus's *Commentary*. This is a

---

[84] *Ibid.*, fols 42r–v: "Proclus, Pappus, Hero, Simplicius, Philoponus, Alexander, Ammonius qui omnes Theonem praecesserunt, nominatim citatis ex Euclide plurimis propositionibus eodem numero quo nunc habentur hanc in Euclidem elementorum eandem referunt ita necessarie, ut eum certe valde caecum esse oporteat et omnis antiquitatis perignarum, qui in ista clarissima luce quamvis minimam de autore dubitationem admittat." The entry in the *Suda*, which Maffei echoed in his biography of Proclus, mentioned that Philoponus wrote a refutation of Proclus's anti-Christian writings. Savile appears to have inferred from this that Proclus and Philoponus (in fact of the late sixth century) were contemporaries.

[85] Bodleian Library, Savile W.12 (*Euclidis Megarensis Geometricorum elementorum*, Paris, 1516. Edited by Henri Étienne) On verso of title-page, in Chamber's hand: "Demonstrationes falso dictitatas Theonis, esse eas Euclidis argumentis sexcentis pervinci potest, nam Proclus, qui ante Theonis tempora florebat, ita harum demonstrationum vestigiis ubique insistit, ut facile constet Euclidem theoremata demonstrationibus non solum illustrasse, sed his ipsis illustrasse. Consulant Proclum prop. 18 ibi *epeidê de ho geômetrês en têi* etc." The passage referred to is at Proclus (1992, pp. 246–247), where Proclus paraphrases the proof of I.18 exactly as it now appears in the Elements, attributing it to "the geometer."

[86] For Proclus, see Frisius (1592, fol. 41r). Frisius also distinguished between Euclid of Megara (fol. 13v, flor. 422 B.C.) and Euclid the geometer (fol. 19r–v, flor. 320 B.C.); he does not mention the confusion between the two men. On Frisius, see (Braun, 1973), pp. 56–57.

paradoxical situation. When almost all of the accepted historical facts around the *Elements* were wrong, it was possible to make arguments both for and against its unity. The false dating of Proclus, in particular, allowed one to argue that Theon had contributed to the *Elements* at least as much as Euclid had (as Ramus thought), or that he had done very little. The supposed existence of two recensions of the text permitted arguments both for and against textual evolution. Now that this chronological error had been resolved, the text resisted analysis both by its critics *and* its supporters; there was nothing upon which to build any sort of historical argument.

Taking into account the revised dating for Proclus, Savile wrote in these late lectures that Euclid provided solid proofs for the results he inherited from his predecessors, a fact established through the testimony of "Proclus, who lived only two centuries after Theon;" thus Euclid was responsible for the *Elements* in its present state. But the argument from Proclus's text was no longer conclusive, as it had been when Proclus was thought to precede Theon. Savile claimed that Proclus's testimony on Euclid's authorship should be trusted because he was sufficiently close to Theon to know the extent of his editorial intervention. Thus the argument became only as strong as one's faith in Proclus as a historical authority.

Aware that the argument from Proclus's text had lost most of its force, Savile cast around for anything that could bolster his claims for the unity of the authorship of the *Elements*. He called those who could imagine any other circumstance "stupid and ridiculous." In addition to questioning the intelligence of his opponents, Savile cited the evidence within the work of Theon himself, who mentioned only once his editing of the *Elements*, and then only with respect to a couple of propositions.[87] He also cited a passage from Alexander of Aphrodisias, an author who, without any doubt, preceded Theon. Alexander quoted the *Elements* in a way that made it clear that his text did not include some propositions found in the modern, post-Theonine text. Savile wished to conclude that these citations demonstrated only *minor* changes to the *Elements*. But all they show is that even the limited evidence still extant from the period attests that Theon *did indeed* make some changes to the text. Aware, it seems, that his arguments fell quite short of certainty, Savile tried a different tack:

> Another obstacle [to the Theonine authorship] is the miraculous, harmonious sequence of the propositions. If you were to remove one proposition from its place, of necessity the entire arrangement and structure would be totally destroyed.

Yet, as we have seen, such a paean to the unity and harmony of the text, derived ultimately from Proclus's *Commentary*, had been made throughout the sixteenth century, most notably by editors who thought that the demonstrations of the *Elements* were the work of Theon! This commonplace argument, which did little to forward his position on the single authorship of the *Elements*, was how he concluded his formal arguments that Theon could have made only the most minimal interventions in the text of the *Elements*.

---

[87] That is, the passage cited from his commentary on the *Almagest* at n. 13 above.

Savile must have known that none of his arguments was conclusive; therefore, his next move was to shift the argument onto new rhetorical and emotional ground. Theon, he said, even if not the author of the *Elements*, was nevertheless an excellent mathematician. He wrote a commentary on the *Almagest*, and had a daughter, Hypatia, who was a distinguished philosopher and mathematician herself. "Even though it is a digression," he told his audience, "perhaps you will enjoy hearing her story."

Over several pages, Savile recounted the story of the beautiful young pagan scientist, who was torn to pieces by a Christian mob. His sympathy lay entirely with her. The trouble began, he says, from her friendship with the prefect Orestes. Orestes had been causing difficulties for the bishop of Alexandria, Cyril ("difficulties," added Savile, "that were entirely justified"). This bishop was Saint Cyril, "a better theologian than human being," who had once also persecuted John Chrysostom. (Savile had reunited Chrysostom's works into a single edition of eight monumental volumes, a task that cost him years of labor and almost brought him to financial ruin. He was inclined to favor Chrysostom in any situation, and hence took an especially dim view of Cyril). Savile went on to recount how Cyril, blaming Hypatia for his problems with Orestes, assembled a gang who intercepted her litter on the way home. They dragged her out of her chair and into a church "where they tore off her clothes, beheaded her, ripped all her limbs off, threw the dismembered body into a pit and set it on fire."[88] Savile drew this description from the church historian Socrates; as if to underline the horror of it, he went on to retell the moment of her death from another source, Hesychius.[89]

After tossing a few more choice insults at Cyril, Savile returned to his sedate and scholarly commentary on the *Elements*. What is the explanation for this gruesome interlude? To some extent, Savile was parading his erudition. He was, as far as I know, the first Renaissance author to relate the story of Hypatia in the context of the history of mathematics.[90] His use of the story in this place in the *Praelectiones* is quite deliberate, and meaningful. Without the false dating of Proclus, Savile was at something of a loss to defend his life-long position on the single authorship of the *Elements*. He could only adduce the logical perfection of the *Elements* in support of his theory: no part of the *Elements* could be separated without destruction of the whole. In the digression that followed on immediately from this argument, Savile

88 Savile (1621, p. 14):"... ubi exutam vestibus testis interficiunt, membratimque discerptam coniciunt in Cinaronem, membraque congesta igne absumunt." Savile explains in note that Cinaro was some kind of garbage dump.

89 *Ibid.*, p. 15: "discerptam scribit ab Alexandrinis, corpusque eius per totam civitatem illusum, propter excellentem eruditionem, maxime circa Astronomiam, paternam haereditatem." ("He writes that she was torn apart by the Alexandrians, and her body defiled through the entire city, because of her great learning inherited from her father, particularly in astronomy.")

90 Ramus, in particular, made no mention of her, for all his interest in her father Theon. The most extensive account of Hypatia, before Savile, was that of Cesare Baronio, in his *Annales ecclesiastici* of 1588–1607. Like Savile, he was quite critical of Cyril's role in the affair. He did not cite Hesychius, and thus was not Savile's source for the story of Hypatia (or, at least, not his only source). See Baronio (1705–1712, vol. 5, pp. 319–320). On Baronio's role in the propagation of the Hypatia story, see Dzielska (1995, p. 23).

dwelt upon the treatment of the body of the virgin Hypatia, as if her body were that of the indivisible *Elements*. The beauty and virtue of the female philosopher, the base character and motives of her assailant, and the visceral horror her slaughter provoked – these, in the end, were the only answer he could find to Ramus, who wished to pull the *Elements* apart into its separately-authored members in precisely such a gruesome fashion.

Savile's juxtaposition of the argument for single authorship and the story of Hypatia was no coincidence. Elsewhere he quite explicitly compared the *Elements* to a human body. Later in the lectures, commenting on the infamous fifth postulate, Savile told his students:

> On the most beautiful body of geometry (*in pulcherrimo Geometriae corpore*) there are two blemishes (*naevi*) and no more, so far as I know. I have spent long hours poring over the writings of ancients and moderns, as you shall soon see, in order to wipe away and erase those blemishes.[91]

In contrast with Ramus, with his revolting fantasies of fingering through the viscera of the *Elements*, Savile (like Proclus) gazed upon a beautiful, almost flawless body. He desired only to complete its perfection, to leave it unblemished, pure, virginal. Hypatia was safe with him. The force of the metaphor is strengthened if we recall that Savile had, through his edition of Chrysostom, already undone the damage to the broken textual body of the theologian, another victim of Cyril's implacable anger.

It is worth considering for a moment what these two *naevi*, blemishes or moles, were on the fair skin of the *Elements*. The first was the fifth postulate itself; the other a flaw in the composition of ratios, which Savile found in the 23rd proposition of the sixth book.[92] These are precisely the blemishes that the unwitting pioneer of non-Euclidean geometry, Girolamo Saccheri, a little over a century later, tried to heal in the two books of his *Euclid freed from every flaw* (*Euclides ex omni naevo vindicatus*"). Saccheri did not mention Savile, yet there can be no doubt that he knew his lectures, which remained a popular introduction to Euclid for some time. It is remarkable to think that the mathematical body that the Jesuit so gallantly defended was, in a sense, that of Hypatia; and that the origins of non-Euclidean geometry are, if tenuously, connected with a long-exploded misunderstanding about the life of Proclus.

In a sense, though, this connection is entirely appropriate. Proclus, misplaced in history, opened up for Ramus the possibility of an internal, historical critique of the *Elements*. He was following a false lead, of course, and his historical criticism was largely nonsense. Genuine historical analysis of the *Elements* would be the province of nineteenth and twentieth-century scholars, from Heiberg and Heath

[91] Savile (1621, p. 140): "In pulcherrimo Geometriae corpore duo sunt naevi, duae labes, nec, quod sciam, plures, in quibus eluendis et emaculandis, cum veterum tum recentiorum, ut postea ostendam, vigilavi industria."

[92] An unpublished manuscript treatise on this latter question by Savile is in the Bodleian Library (MS Savile 108, fols 59r–66r) and the Ambrosiana Library, Milan (MS D.243 inf., fols 1r–6v).

to Knorr and Fowler, based partly on manuscript evidence unknown in the Renaissance.[93] Nevertheless, Ramus suggested a new way of reading the *Elements*, as a historical palimpsest – a mode of reading that had never been considered before.

In response, Savile revived the notion of the *Elements* as an ahistorical text, perfectly logical and indivisible (at least, indivisible without violence). He expressed this unity through the metaphor of a female body, blemished only by a couple of moles – a tender, even eroticized image. Saccheri's defense of Euclid similarly ignored the historicity of the text. For him, like Savile, the *Elements* was a single body on which, with the scalpel of the logician, he performed delicate cosmetic surgery to cure its only flaws. His failure was, of course, momentous. After him, the logical limitations of the *Elements* and the contingency of its postulates would be evident – and the *Elements* would once again be open to historical analysis.

[93] In particular, the tenth-century manuscript P (MS Vatican Cod. Gr. 190), which lacked the customary statement that it was from the edition of Theon, and which also did not contain the addition to proposition VI.33, which Theon claimed to have written. This demonstrates a pre-Theonine basis to the manuscript (though the copyist seems also to have used post-Theonine manuscripts at times). See Euclid (1926, vol. 1, pp. 46–63); Heiberg (1882, pp. 174–180).

# Conclusion

Savile died a year after the publication of his Euclid lectures. He could be sure that the future of mathematics at Oxford was secure. He had established his two professorships of mathematics and seen them filled with his hand-picked successors, Henry Briggs and John Bainbridge. What is more, he provided his professors with a large library of printed books and manuscripts he had collected over the course of his career, as well as his own personal working papers, including the volumes of his 1570 lectures. The professors were also provided with instruments and a handsomely endowed mathematical "chest" for funding their teaching and research.

With this foundation, Savile sought to remedy the university's long neglect of mathematics, as he had described it in his 1570 lecture. He also guaranteed a place at Oxford after he had gone for the kind of humanistically-grounded mathematics that he himself had pursued, although tempered now by a degree of utility.

But just as Savile had, in his very first mathematical enterprise, patterned his efforts on Ramus, so in his final efforts to ensure his legacy within the academy, Savile was following in the footsteps of his French model and rival. For Ramus, too, had left a will with instructions providing for the future of his mathematical program. Ramus had been revising his *Prooemium mathematicum* for a third time in 1572 when he was caught up in the horror of the Saint Bartholomew's Day Massacre. The Catholic mob (rumored to include Charpentier himself) shot the Protestant professor, then ran him through with a sword, threw him from the window of his study and then, finding him still alive on the pavement below, drowned him in the Seine. His library was looted and his papers scattered.

Ramus's posthumous fortunes were complex. Among Protestants abroad, his reputation soared. Johann Freig, who popularized his work by means of the dichotomous tables that have now become synonymous with "Ramism," provides perhaps the most extraordinary example of how hyperbolic the veneration of Ramus could become. At the end of the hagiographic biography of Ramus that he appended to an edition of his works, Freig related how "Christ the Lord and our Savior" healed a blind man by rubbing in his eyes a mixture of dirt and spittle. "But why am I telling you this?" asked Freig rhetorically: "For me, who had been blind in philosophy for many years, Ramus was like Christ."[1]

[1] Ramus (1599, p. 612): "Mihi, inquam, in philosophia multos annos caeco Ramus instar Christi fuit."

Slightly more tempered praise was found in John Chamber's 1575 ordinary lectures on astronomy at Oxford. Ramus was, he said, "not only a great mathematician, but also very learned otherwise"; elsewhere he called him "a man of remarkable merit, not quite born, but rather given to us by God Himself, whom we all very rightly ought to love, on account of his great holiness."[2] What is remarkable about Chamber's praise of Ramus is that his lectures were largely copied from those that his friend Henry Savile delivered in 1570; but of course Savile, in his lectures (written before Ramus's death) had attacked Ramus quite mercilessly, both for his style of mathematics and the tenor of his history. Perhaps it was simply tasteless to attack Ramus's reputation after his death.

But Peter Ramus, Protestant martyr, had become so closely associated with mathematics that it would have been counterproductive in any case for a promoter of mathematics to say anything too critical of its best-known representative. For, although Ramus never held a chair of mathematics in Paris, he was very soon assumed to have done so. Many works written after his death refer to him as "Regius Professor of mathematics." On the title-page of Thomas Hood's 1590 English version of Ramus's *Geometry*, for instance, he is described as "that excellent scholler P. Ramus, Professor of the mathematical sciences in the University of Paris."[3]

Ramus's legacy, though assured for some time in Protestant Europe, was bitterly contested in Paris. He had left a will, in which provisions were made for the foundation of a mathematical professorship in Paris. In order to exert the control over the teaching of mathematics that had eluded him in the last years of his life, Ramus hedged the appointment about with conditions. The professor was to teach arithmetic, music, geometry, optics, mechanics, geography and astronomy "and was to do so not according to the opinion of men, but according to reason and truth."[4] Ramus intended the first professor to be Friedrich Risner, best known today for his 1572 edition of the optical works of Witelo and Alhazen, which were the starting point of Kepler's optical investigations; this textual work was the fruit of his long collaboration with Ramus, as was an original work on optics that Risner later published under their joint names.[5] Risner was, said Ramus, to hold the chair for 3 years, during which time he must teach all the mathematical subjects in the manner Ramus had specified. If he did so successfully, he could be appointed for a further 3 years. If he did not fulfill the terms of the will, then a new professor would have to be chosen, according to a method that Ramus set out in detail in this document.

At last, then, Ramus was able to impose an examination on the holder of this chair of just the sort that he had wished vainly to require of Charpentier. Three months before the intended examination, any prospective candidates would be invited to teach at Paris in order to demonstrate their expertise. They would only be admitted

---

[2] Chamber (1601, *Astronomiae encomium*, p. 14): "a Petro Ramo non mathematico solum magno, verum etiam caeteroque perdocto viro"; p. 6: "Petrus Ramus, homo ad laudem insignis non omnino natus, sed ab ipso deo affectus, quem omnes amare meritissimo pro eius eximia sanctitate."

[3] Ramus (1590).

[4] The text of the will is in Waddington (1855, pp. 326–328).

[5] Risner (1572, 1606).

to the competition if they were literate in Latin and Greek, as well as in all the non-mathematical liberal arts. Then the examination would take place, before the body of regius professors, the President of the Paris Parlement and other dignitaries. On each of the first 7 days of the exam, the candidates would each deliver an hour-long lecture on each of seven mathematical sciences that Ramus had listed. On the eighth, they were required to answer questions and solve problems, including providing a proof on the spot of any theorem offered by a member of the audience. (This last detail recalled the moment in Ramus's battle with Charpentier before Parlement, in which Ramus dramatically produced a copy of the Elements from his robe, saying that he would drop his entire suit against Charpentier if he could prove then and there just one theorem from the book; Charpentier, naturally, did not take up his offer.) The candidate judged both the best mathematician and the best teacher would be appointed for 3 years, at the end of which, regardless of how well he had discharged his office, there would be another public examination, in which he would have to defend his position against all comers. And so it would proceed every 3 years.

It is remarkable the degree to which Ramus wanted his professors to be made in his own image. His stipulation that they must teach mathematics "not according to the opinion of men, but according to reason and truth" was meant to imply that they were to teach according to his method – perhaps even from the textbooks he had prepared on the mathematical sciences. For Ramus believed that it was a distinguishing mark of his arithmetic and geometry that he had finally restored these sciences to their original rational order, so that the truth of their propositions shone forth without the need for the man-made demonstrations of Euclid and other ancients and moderns. He even required that the professor's first lecture should be "an encomium of mathematics, whereby he exhorts the youth to its study" – a not unreasonable way to begin a series of lectures; but it is also not insignificant, I think, that this was exactly the way that Ramus had begun his own lectures on mathematics.

Ramus's checkered reputation in Paris ensured that his legacy would not fare smoothly: neither the professorship nor his program of mathematical reform. Immediately after the reading of the will, the regius professors of the Collège petitioned Parlement, saying that there was no need for a mathematical professor and the money could be better spent elsewhere. Parlement agreed, and assigned the monies to the doctor, litérateur and dabbler in alchemy Jacques Gohorry, for the writing of his history of France.[6] This led, inevitably, to protests from Ramus's friends, the most compelling of which was made in 1575 upon the death of Gohorry by the executors of the will, Nicholas Bergeron and Antoine Loysel. They convinced Parlement that the funds needed to be used for their intended purpose, and that Risner should be summoned to fulfill his part in Ramus's last wishes. Yet this turned out very poorly, as Bergeron recorded in pamphlets he published in 1576 and 1580 in order to set out publicly the executors' side of the affair.[7] The executors had sent Risner an advance on his salary, and brought him to Paris to teach. When he arrived

[6] See Waddington (1855, pp. 334–338), on the obstruction to the will.

[7] Bergeron (1576, 1580).

"by some strange impulse, or maybe even by mental illness" he refused to deliver a single lecture, let alone take on the responsibilities of a professor. After his return to Germany, the executors asked for the return of their money; but Risner had already spent all of it, he said, on research into the mathematical sciences which would be published "in accordance with the intention of the deceased," as he claimed.[8] Risner's last association with Ramus had been as a fellow researcher and writer – that was the Ramist legacy that he intended to preserve.

Bergeron's first pamphlet, written directly after the debacle with Risner, opened with an engraving of Ramus's portrait, beneath which was a poem by Bergeron himself that began: "Ramus, who once lay mutilated, wounded, stabbed beneath the dreadful waves now has sprung forth again from the foul earth!" Another dedicatory poem (also by Bergeron) opened: "Henceforth the golden bough (*ramus*) begins to green from the sacred tree, bearing golden fruit for the rest of time."[9] Ramus would be brought back to life by honoring the terms of the will. To drive this point home, a part of the pamphlet consisted of the text of the will itself – printed especially because Bergeron found that few knew what Ramus had actually stipulated about the professorship. What is more, Bergeron as executor demanded that engraved copies of the will be affixed at several places in the University (including in Ramus's own college, the Collège de Presles, as if to assert his continuing presence there).[10] Since Risner was clearly not going to take up his duties, a new competition for the professorship should be advertised. The terms of the competition, as Bergeron laid them out, were exactly as Ramus had required, even stating that the successful professor would have to lecture for 3 years "according to Ramus's method."

The competition was held and a professor, Maurice Bressieu, duly chosen. Throughout the period of the examination, the passage from Ramus's will concerning the professorship, engraved on a brass plate, was displayed in the places where the disputations took place.[11] The ostensible aim was to remind all the participants of the terms of the competition; but Ramus's resurrected words were, in a sense, witnesses to the actions they had set in motion. Ramus's posthumous intentions were made to stand very concretely for his continuing influence as an actor in the University. Ramus had retold the biography of Pythagoras so as to make him seem a Ramist schoolmaster. And now, at last, the University of Paris was Pythagorean once more.

In fact, however, it was his writings that were to have a more lasting influence on European mathematics. His chair would only be held by one mathematician of any prominence, Gilles Roberval (1602–1675). In Oxford, however, in the century after Savile's death, the Savilian professors provided for the mathematical education

[8] On Risner's spending of the Ramus money on research, see Bergeron (1576, pp. 11–12); on his strange behavior in Paris, see Bergeron (1580, p. 18).

[9] Bergeron (1576, p. 2): "Qui iacuit miseris mutilus lacer obrutus undis / Ramus, ab obscoena iam revirescit humo!" and "Aureus hinc sacra frondescit ab arbore Ramus / Aurea perpetuo tempore poma ferens."

[10] *Ibid.*, p. 11.

[11] Bergeron (1580, p. 19).

of generations of Oxford undergraduates while also doing much to further the state of mathematical and astronomical research in England. Among those who held the chairs were leading scholars and practitioners like John Wallis, Christopher Wren, Edmund Halley, and David Gregory, whose monumental edition of the works of Euclid (1703) furthered Savile's editorial project and even relied on his manuscripts and notes.

A decade later, de Montmort wrote his letter to Bernoulli, musing on the possibility of writing a history of mathematics, with which I began this book. These Enlightenment *savants* marveled at the advanced state of mathematics in their time and wished for a history that would do it justice. They had no doubt that mathematics was a noble art, standing at the very pinnacle of human intellectual endeavor. It was thanks to Savile and Ramus's efforts, in part at least, that mathematics had progressed so far. Some irony, then, in the fact that their extensive, learned, protreptic exhortations to the cultivation of the sciences were now not only no longer needed, but in fact long-forgotten. De Montmort, Bernoulli and, ultimately, Montucla, imagined themselves to be the first to even consider the question of the origins of the mathematical art. Yet they were, in fact, the beneficiaries of Renaissance humanists who each, in his own way, had labored to build a vision of Alexandria in his native land – and who wrote the future of mathematics just as he had written its past.

# Appendix A
# Contents of Savile's History of Mathematics

**(Bodleian Library MS Savile 29)**

| Fol. | Heading in Manuscript |
|---|---|
| 29r | Sethi filii |
| 29r | Noä (in margin: Seth, Adam, Cain, Armenii) |
| 29v | Samothes. Druides |
| 29v | Magi. Zoroastres |
| 29v–30r | Chaldaei |
| 30r–31r | Abraham. Phoenices. Aegyptii |
| 31r | Joseph. Albion. Jacob Levi. Theuth |
| 31r-v | Atlas. Trismegistus |
| 31v–32r | Hyas. Orion. Aristeus. Melampus. Phineus. Chiron. Homerus. Hesiodus. Argonautae. Berosus |
| 32r-v | Thales. |
| 32v–33r | Ameristus. Pythagoras |
| 33r-v | Anaximander. Anaximenes |
| 33v–34r | Anaxagoras. Oenopides. Hippocrates. Democritus. Bion |
| 34r | Leostratus. Meton. Euctemon |
| 34v–35r | Hippocrates Chius. Briso. Antipho |
| 35v | Theodorus Cyrenaeus |
| 35v–36v | Archytas. Philolaus. Nicetas. Heraclides. Ecphant. Eudoxus |
| 36v–37r | Plato. Leodamas. Theaetetus. Nioclides. Leon |
| 37v–38r | Amyclas. Menaechinus. Dinostrates. Theudius. Cizicenus. Hermotimus. Phillipus. ~~Heraclides Ponticus~~. Xenocrates. Speusippus. Amphinomus |
| 38r | Aethiopes. Aegyptii, Libyes, Babylonii, Orphei, Lyra Tiresias Atreus et Thyestes. Bellerophon. Phrixus. Daedalus. Icarus. Pasiphaë. Endymion. Phaëthon |
| 41r–44v | Euclides |

| | |
|---|---|
| 44v–45r | Aristoteles. Theophrastus. Aristoxenus. Dicaearchus. Eudemus. Calippus |
| 45r–46r | Cynici. Stoici. Epicurei |
| 46r–47r | Autolycus. Hipponicus. Arcesilas. Lacydes. Aristyllus. Timocharis. Diognetus. Callias. Aratus. Aristarchus. Dionysius |
| 47r–51v | Archimedes |
| 52r | Eudoxus. Panaetius. Archelaus. Cassandrus. Scylax Italicus.[1] |
| 56r | Conon. Dositheus. Eratosthenes. Apollonius Rhodius |
| 56r–57v | Apollonius Pergaeus. Hipparchus |
| 58r | Dionysodorus. Aeneas. Hippias. Perseus. Aristaeus. Xenodotus. Serenus |
| 58r–59r | Ctesibius. Hero. Geminus. Carpus. Zenodorus |
| 59r | Diophantus. Nicomachus |
| 59r-v | Philo. Diocles. Sporus. Nicomedes |
| 59v–60r | Posidonius. Panaetius. Theodosius |
| 60r–61v | Caesar. Sosigenes. Taruntius. Cicero. Manilius. Higinus. Vitruvius. Virgilius. Germanicus. Plinius. Meto. Solinus |
| 61v–62v | Apollonius. Dionysius Areopagita. Strabo. Dionysius Afer. Maximus. Cleomedes. Andromachus. Agrippa. Menelaus. Dionysius Halicarnassus. Theon |
| 62v–65v | Ptolemaeus |

[1] This heading appears on a blank page, with a reference to Cicero's *De divinatione*.

# Appendix B
# Evidence for the Extent of Savile's Lectures

Anthony à Wood marvelled at the accomplishment of Savile's 1570 lectures[1]; without a doubt they were an important moment in the teaching of astronomy in England. But the composition of the lectures and the circumstances of their delivery have been little studied and frequently mischaracterized. It is not clear that Savile was ever able to deliver the complete set of lectures as they are extant in manuscript. The manuscripts, we recall, contain the following material:

1. Protreptic exordium (MS Savile 29, fols 2r–8r).
2. Introduction to the seven mathematical sciences (according to the classification of Geminus), arithmetic, geometry, music, optics, mechanics, astronomy and geography (MS Savile 29, fols 8r–25r).
3. History of mathematics from Adam to Ptolemy (MS Savile 29, fols 29r–65v).
4. Commentary on the mathematics of the *Almagest* (remainder of MS Savile 29, fols 65v–140r, and MSS Savile 31 and 32).

At the head of the first lecture in MS Savile 29, notes, Savile wrote the date "10 October 1570," the beginning of Michaelmas term.[2] The first 65 folios of this volume (Sects. 1–3 as I have numbered them above) are not clearly divided into individual lectures, but it is possible to infer that they represent eight lectures in total. At the end of his lecture on arithmetic (immediately following the protreptic) Savile wrote (in Greek) "*telos*" and, in the last few sentences, told his students that he was unwilling to prolong this lecture late into the day. This is the first indication in the lectures that he has reached a break, so we can assume that the protreptic together with the description of arithmetic (about 8,000 words in total) together constitute the first lecture Savile actually delivered. In the same passage, Savile also said that it would take him two further lectures to cover the remaining sciences; and the mate-

[1] See p. 75 above. Wood may have heard of the lectures from John Wallis who, as Savilian Professor, would have kept the three manuscripts in his study along with the rest of Savile's library. He may also have known the meticulous description of Savile's manuscript remains that was prepared by the antiquarian Gerard Langbaine (1609–1658; this description is now MS Savile 107); many of Langbaine's bibliographical papers now in the Bodleian Library were once in Wood's possession. See the Bodleian Library *Summary Catalogue* vol. 1, pp. xviii–xxv.

[2] MS Savile 29, fol. 2r.

rial on the six remaining arts does indeed extend to about 16,000 words.[3] The third section of the lectures, Savile's historical excursus, amounts to some 40,000 words and so must have been delivered over about five lectures.

In the fourth section, the exposition of Ptolemy's *Almagest*, Savile began to mark the divisions between individual lectures, numbering them from 1 to 37.[4] At the beginning of the lecture following the 37th, he wrote "First lecture, new term."[5] He continued to divide his material into lectures, but less conscientiously than before, and by the end of the volume had abandoned this practice altogether. It is clear enough, however, that MS Savile 29 contains, in total, the notes for two terms of lectures.

Until the early twentieth century, there were four terms in the Oxford academic year (Michaelmas, Hilary, Easter and Trinity);[6] the natural inference would be that MS Savile 29 contains lecture notes for Michaelmas and Hilary, 1570–1571. We should expect, then, to find in MS Savile 31 lecture notes for Easter and Trinity terms of the same academic year. But it seems that Savile did not lecture continuously through 1570–1571. The University was sparsely attended that spring because of the threat of plague. All members of Merton College were granted liberty to leave. The *Register* of Merton College records that, on April 13, 1571, the Vice-Warden of the college and the senior Fellows met to consider the predicament of the ordinary lecturers whom the University still required to deliver another two terms of lectures that year. They decided that Savile and three other masters should be relieved of their teaching obligations for Easter term *simpliciter* (that is, entirely released with no requirement to make up the missed classes). In the final, Trinity term, they would also be released *simpliciter* if the university remained empty for the entire term; if it was open for some of the term, they should make every effort to deliver an entire term of lectures during the period of residence.[7]

In other words, for the year 1570–1571 Savile delivered at most three terms of lectures, and perhaps as few as two. The following year seems to have been disrupted as well. At the beginning of Michaelmas term, 1571, when Savile should have been starting the second year of his ordinary lectures, the Warden and senior

---

[3] MS Savile 29, fol. 10r: "Reliquae mihi sunt ad mathesin expetendam geometria, musica, perspectiva, mechanica et Astronomia, de quibus similiter dicendum est, sed duobus, ut arbitror, secundùm hunc diem proximis. Nam et hora me vocat, et sermonem in multum diem produci nolim. In crastinum."

[4] Each of these strictly mathematical lectures is much shorter than the 8,000 word introductory and historical lectures. They generally consist of diagrams and abbreviated mathematical notes that Savile must have expanded upon greatly in class.

[5] *Ibid*, fol. 109r: "Novus terminus Lect. 1a." It should be noted that the Library's *Summary Catalogue* records this as "a new series for the ninth term at fol. 109," misreading "novus" as "nonus." This reading has been accepted by most writers, with the consequence that Savile's lectures have been thought to have lasted much longer than they probably did – certainly far beyond the required extent for ordinary lectures. See, for example, Feingold (1984, p. 47), where Savile's lectures are said to have lasted from 1570 to 1575; an error repeated in Goulding (1999).

[6] Gibson (1931, pp. lxxxi, 343, 355).

[7] Fletcher (1976, p. 36).

Fellows again permitted Savile and the three other ordinary lecturers to abandon their lectures.[8] Perhaps there was still some fear of the plague; or perhaps the four regent masters whose first year of teaching had been disrupted were finding it difficult to resume their courses of lectures after so long a hiatus. Nevertheless, Savile may not have accepted this second dispensation. In the same year, 1571, a university committee gave notice that some of the previous year's regents would have to stay on in the coming year in order to make up for a shortfall in lecturers. The decree mentions no names, but does require the lecturer in astronomy to continue; and it seems highly likely that this refers to Savile.[9] However, since we have no way of knowing whether the university decree was intended to overrule the dispensation Merton College had offered Savile, or vice versa, we cannot say for certain whether he lectured at all in the year 1571–1572.[10]

What is clear is that Savile delivered his protreptic, his introduction to the individual mathematical arts and his historical excursus in the course of his ordinary lectures for the Michaelmas term of 1570–1571. In the same term he also gave 37 lectures on the *Almagest* itself, and continued into the Hilary term without interruption. Here, he covered only the mathematical and geometrical preliminaries to the study of the *Almagest* and elementary spherical astronomy. It is less certain whether he went on actually to deliver the advanced commentaries on the *Almagest* that are contained in MSS Savile 31 and 32, which cover planetary astronomy, Copernicus and much else at a very high level. Thus, it may well be that Savile's fame derived purely from the protreptic, historical and elementary mathematical sections of the work.

[8] *Ibid.*, p. 39. It may be that there was an expectation that they would complete their regency requirements some time in the future, since they were also permitted by this decree to perform their "variation" (the public disputations that marked the end of the arts regency) any time in the next 3 years.

[9] Clark (1887–1889, I, p. 98).

[10] It may also be worth noting that, when he started his lectures in 1570, Savile seems to have had the intention of delivering only *one* year of lectures. At MS Savile 29, fols 7v–8r, Savile says that he will not make excessive claims of how much astronomy the students can learn, curbed as he is by "the limits of my province, to which I have been sent for a year with consular powers" ("meae provinciae in quam ad annum missus sum cum imperio, limitibus concludatur").

# References

## Primary Texts


Alfraganus, *Rudimenta astronomica*, Nuremberg, 1537. Prefaced by Regiomontanus's oration on the history of mathematics.

Joannes Annius, *Berosi sacerdotis chaldaici antiquitatum libri quinque*, Antwerp, 1545.

Anonymous, *In Petri Rami insolentissimum Decanatum ... Philippica prima*, Paris, 1567. Published anonymously, although clearly by a pupil of Charpentier's (if not Charpentier himself).

Aristippus, *Aristippi et Cyrenaicorum fragmenta*, ed. E. Mannebach, Leiden: E. J. Brill, 1961.

John Aubrey, *Brief Lives*, ed. O. L. Dick, London: Secker & Warburg, 1958. 3rd ed.

Cesare Baronio, *Annales ecclesiastici*, Venice, 1705–12.

Nicolas Bergeron, *Testamentum Petri Rami cum senatusconsulto et promulgatione professionis institutae ab ipso testatore*, Paris, 1576.

Nicolas Bergeron, *Commendatio professionis mathematicae, a Petro Ramo institutae, ingratiam Academiae Parisiensis*, Paris, 1580. With French translation by the author.

Jean Borrel, *De quadratura circuli libri duo, ubi multorum quadraturae confutantur, et ab omni impugnatione defenditur Archimedes*, Lyon, 1559.

Giordano Bruno, "De la causa, principio et uno" in *Œuvres complètes*, ed. Giovanni Aquilecchia, 3, Paris: Les Belles Lettres, 1996.

Campanus, *Campanus of Novara and Euclid's elements*, ed. H. L. L. Busard, *Boethius* 51, Stuttgart: Franz Steiner Verlag, 2005.

Girolamo Cardano, *Opera omnia*, Lyon, 1663. 10 vols.

John Chamber, *A Treatise Against Iudicial Astrologie; Astronomiae Encomium*, London, 1601.

Jacques Charpentier, *Ad expositionem disputationis de methodo, contra Thessalum Ossatum, Academiae Parisiensis Methodicum, Responsio*, Paris, 1564.

Jacques Charpentier, *Orationes tres, pro iure professionis suae, in Senatu extempore habitae, contra importunas Rami actiones*, Paris, 1566.

Jacques Charpentier, *Admonitio ad Thessalum, Academiae Parisiensis methodicum: de aliquot capitibus Prooemii mathematici, quae continent eiusdem Carpentarii Praelectiones in sphaeram*, Paris, 1567.

John Dillon and Lloyd P. Gerson, *Neoplatonic Philosophy: Introductory Readings*, Indianapolis: Hackett 2004.

Euclid, *Euclidis Megarensis philosophi platonici ... habent in hoc volumine quicunque ad mathematicam substantiam aspirat: Elementorum libros XIII ...*, ed. and tr. Bartolomeo Zamberti, Venice: Ioannes Tacuinus, 1505.

Euclid, *Euclidis Megarensis Geometricorum elementorum XV*, ed. Jacques Lefèvre D'Étaples, Paris, 1516.

Euclid, *Eukleidou stoicheiôn bibl. 15 ek tôn Theonos synousiôn*, ed. Simon Grynaeus, Basel, 1533. (Also Greek text of Proclus's *Commentary*.)

Euclid, *Euclidis Megarensis mathematici clarissimi Elementorum geometricorum libri XV*, ed. Johannes Hervagius, Basel, 1537. (Reprinted by Hervagius in Basel in 1546.)

Euclid, *In sex priores libros geometricorum elementorum Euclidis demonstrationes, una cum ipsius Euclidis textu graeco, et interpretatione latina B. Zamberti*, ed. Oronce Fine, Paris, 1544.

Euclid, *Elementa*, ed. P. Ramus, Paris, 1549.

Euclid, *In Euclidis Elementa Geometrica demonstrationum libri sex*, ed. Jacques Peletier, Lyon, 1557.

Euclid, *Euclidis Megarensis mathematici clarissimi Elementa geometrica*, ed. François de Foix de Candalle, Paris, 1566.

Euclid, *Elementorum libri XV*, ed. Federico Commandino, Pesaro, 1572.

Euclid, *The Elements of Euclid, viz., the first six books together with the eleventh and the twelfth; the errors, by which Theon, or others, have long ago vitiated these books, are corrected...*, ed. Robert Simson, Edinburgh, 1775.

Euclid, *The Thirteen Books of the Elements*, ed. Thomas L. Heath, Cambridge: Cambridge University Press, 1926. 2nd ed. 3 vols.

Marsilio Ficino, *Epistolae*, Nuremberg, 1497.

J. M. Fletcher, *Registrum annalium Collegii Mertonensis 1567–1603*, Oxford: Clarendon, 1976.

Jean-Jacques Frisius, *Bibliotheca philosophorum classicorum authorum chronologica*, Zurich, 1592.

Konrad Gesner, *Bibliotheca universalis*, Zurich, 1545.

Strickland Gibson, *Statuta antiqua Universitatis Oxoniensis*, Oxford: Clarendon, 1931.

Raffaele Maffei, *Commentariorum urbanorum octo et triginta libri*, Rome, 1506.

Jean-Papire Masson, *Elogia varia*, Paris, 1638. Divided into two, separately paginated volumes. *Vita Iacobi Carpentarii* (originally Paris, 1574) at vol. 2, pp. 269–74.

Francesco Maurolyco, *Sicanicarum rerum compendium*, Messana, 1562.

Philip Melanchthon, *Mathematicarum disciplinarum tum etiam astrologiae encomia*, Strassburg, 1537.

Philip Melanchthon, *Philippi Melanthonis Operaquae supersunt omnia, Corpus reformatorum* 1–28, ed. C. G. Brettschneider and H. E. Bindseil, Halle, 1834–1860.

Philip Melanchthon, *Orations on Philosophy and Education*, ed. and tr. Sachiko Kusukawa, Cambridge: Cambridge University Press, 1999.

Pierre de Mondoré, *Euclidis Elementorum liber decimus*, Paris, 1552.

Caspar Peucer, *Elementa doctrinae de circulis coelestibus et primo motu*, Wittenberg, 1553.

Johannes Philoponus, *Commentaria in libros Posteriorum Aristotelis*, tr. Andreas Gratiolus and Phillipus Theodosius, Venice, 1542.

Plato, *Opera*, tr. Marsilio Ficino, Venice, 1491.

Plotinus, *Opera*, tr. Marsilio Ficino, Florence, 1492.

Plutarch, *Vite Plutarchi Cheronei post Pyladen Brixianum longe diligentius reposite*, Paris, 1514.

Proclus, *In primum Euclidis Elementorum librum commentariorum ... libri IIII*, tr. Francesco Barozzi, Padua, 1560.

Proclus, *In primum Euclidis Elementorum librum commentarii*, ed. Gottfried Friedlein, Leipzig: B. G. Teubner, 1873.

Proclus, *A Commentary on the First Book of Euclid's* Elements, tr. Glenn R. Morrow, Princeton: Princeton University Press, 1992.

Petrus Ramus, *Dialecticae institutiones, ad celeberrimam, et illustrissimam Lutetiae Parisiorum Academiam*, Paris, 1543a.

Petrus Ramus, *Aristotelicae animadversiones*, Paris, 1543b.

Petrus Ramus, *Brutinae quaestiones*, Paris, 1549.

Petrus Ramus, *De moribus veterum Gallorum*, Paris, 1559.

Petrus Ramus, *Prooemium mathematicum*, Paris, 1567.

Petrus Ramus, *Scholarum mathematicarum libri unus et triginta*, Basel, 1569.

Petrus Ramus, *De religione christiana libri quatuor*, Frankfurt a. M., 1576.

Petrus Ramus, *Elementes of geometrie*, tr. Thomas Hood, London, 1590.

Petrus Ramus and Audomarus Talaeus, *Collectaneae praefationes, epistolae, orationes*, Marburg, 1599.
Johannes Regiomontanus, *In Ptolemaei magnam compositionem ... Epitome*, Nuremberg, 1550.
Johannes Regiomontanus, *Opera collectanea: Faksimiledrucke von neun Schriften Regiomontans*, Osnabrück: O. Zeller, 1972.
Friedrich Risner, *Opticae thesaurus*, Basel, 1572.
Friedrich Risner and Petrus Ramus, *Opticae libri quatuor ex voto Petri Rami novissimo*, Kassel, 1606.
Henry Savile, *Praelectiones tresdecim in principium Elementorum Euclidis Oxonii habitae MDCXX*, Oxford, 1621.
Johannes Stöffler, *In Procli Sphaeram*, Tübingen, 1534.
Theon and Pappus, *Commentaires de Pappus et de Théon d'Alexandrie sur l'Almageste*, ed. A. Rome, Vatican City, 1936.
Polydore Vergil, *On Discovery*, ed. and tr. Brian Copenhaver, *I Tatti Renaissance Library*, Cambridge, MA: Harvard University Press, 2002.
Johannes Vögelin, *Elementa geometriae*, Wittenberg, 1536.
Anthony à Wood, *Athenae Oxonienses. An Exact History of All the Writers and Bishops Who Have Had Their Education in the University of Oxford*, ed. P. Bliss, London, 1813–20. 4 vols.
Gulielmus Xylander, *M. Antonii imperatoris De seipso seu vita; Marini Neapolitani de Procli vita et foelicitate*, Zürich, 1559.

## Secondary Texts

Michael J. B. Allen, *Synoptic Art: Marsilio Ficino on the History of Platonic Interpretation*, Istituto Nazionale di Studi sul Rinascimento: *Studi e Testi* 40, Florence: Leo S. Olschki, 1998.
Giovanni Aquilecchia, "Ancora su Giordano Bruno ad Oxford (in margine ad una recente segnalazione)," *Studi secenteschi* 4 (1963), pp. 3–13.
Giovanni Aquilecchia, "Giordano Bruno in Inghilterra (1583–1585): Documenti e testimonianze," *Bruniana e Campanelliana* 1 (1995), pp. 21–42.
Benno Artmann, *Euclid: The Creation of Mathematics*, New York: Springer-Verlag, 1999.
Eric Ash, *Power, Knowledge and Expertise in Elizabethan England*, Baltimore: Johns Hopkins University Press, 2004.
Dale B. Billingsley, "Authority in Early Editions of Euclid's *Elements*," *Fifteenth-Century Studies* 20 (1993), pp. 1–14.
Ann Blair, "Reading Strategies for Coping with Information Overload ca. 1550–1700," *Journal of the History of Ideas* 64 (2003), pp. 11–28.
Lucien Braun, *Histoire de l'histoire de la philosophie, Associations des publications près les Universités de Strasbourg* 150, Paris: Editions Ophrys, 1973.
Nelly Bruyère, *Méthode et dialectique dans l'œuvre de La Ramée: Renaissance et âge classique, De Pétrarque à Descartes* 45, Paris: J.Vrin, 1984.
James Byrne, "A Humanist History of Mathematics?: Regiomontanus's Padua Oration in Context," *Journal of the History of Ideas* 67 (2006), pp. 41–61.
Andrew Clark, *Register of the University of Oxford, Oxford Historical Society* 1, 10–12, 14, Oxford: Oxford Historical Society, 1887–9 (in 2 vols, the second of which is in 4 parts).
Brian P. Copenhaver and Charles B. Schmitt, *Renaissance Philosophy, A History of Western Philosophy* 3, Oxford: Oxford University Press, 1992.
Serafina Cuomo, *Ancient Mathematics*, London: Routledge, 2001.
Mark Curtis, *Oxford and Cambridge in Transition, 1558–1642: An Essay on Changing Relations Between the English universities and English society*, Oxford: Clarendon Press, 1959.
Augustus de Morgan, "Eucleides" in *Dictionary of Greek and Roman Biography and Mythology*, ed. William George Smith, pp. 63b–74b, Boston: C. Little and J. Brown, 1870.

Maria Dzielska, *Hypatia of Alexandria*, tr. F. Lyra, Cambridge, MA: Harvard University Press, 1995.

Graziella Federici Vescovini, "La Vita dei matematici arabi di Bernardino Baldi" in *Storici, filosofi e cultura umanistica a Gubbio tra Cinque e Seicento: Atti del convegno di studi, Gubbio 6–8 aprile 1995*, ed. Patrizia Castelli and Giancarlo Pellegrini, pp. 57–72, Spoleto: Centro italiano di studi sull'Alto Medioevo, 1998.

Mordechai Feingold, *The Mathematicians' Apprenticeship: Science, Universities and Society in England, 1560–1640*, Cambridge and New York: Cambridge University Press, 1984.

Mordechai Feingold, "English Ramism: A Reinterpretation" in *The Influence of Petrus Ramus: Studies in Sixteenth and Seventeenth Century Philosophy and Science*, ed. Mordechai Feingold, Joseph S. Freedman, and Wolfgang Rother, *Schwabe Philosophica* 1, pp. 127–76, Basel: Schwabe, 2001.

Louis H. Feldman, "Abraham the Greek philosopher in Josephus," *Transactions and Proceedings of the American Philological Association* 99 (1968), pp. 143–56.

J. M. Fletcher, "The Faculty of Arts" in *The History of the University of Oxford*, ed. J. McConica, 3, pp. 157–99, Oxford: Oxford University Press, 1986.

Owen Gingerich, "The Role of Erasmus Reinhold and the Prutenic Tables in the Dissemination of the Copernican Theory," *Studia Copernicana* 6 (1973), pp. 43–62.

Jean-Eudes Girot, "La notion de lecteur royal: le cas de René Guillon (1500–1570)" in *Les origines du Collège de France (1500–1560): actes du colloque international (Paris, décembre 1995)*, ed. Marc Fumaroli, pp. 43–108, Paris: Klincksieck, 1998.

Robert Goulding, "Henry Savile and the Tychonic World-System," *Journal of the Warburg and Courtauld Institutes* 58 (1995), pp. 152–79.

Robert Goulding, "Testimonia Humanitatis: The Early Lectures of Henry Savile" in *Sir Thomas Gresham and Gresham College: Studies in the intellectual history of London in the sixteenth and seventeenth centuries*, ed. Francis Ames-Lewis, pp. 125–45, Aldershot: Ashgate, 1999.

Robert Goulding, "Science at the Elizabethan Universities" in *Reassessing Tudor Humanism*, ed. Jonathan Woolfson, pp. 223–42, Basingstoke and New York: Palgrave Macmillan, 2002.

Robert Goulding, "Henry Savile" in *Oxford Dictionary of National Biography*, ed. H. C. G. Matthew and Brian Harrison, Oxford: Oxford University Press, 2004a.

Robert Goulding, "Henry Savile" in *The Dictionary of British Classicists*, ed. Robert B. Todd, Bristol: Thoemmes Continuum, 2004b.

Robert Goulding, "Polemic in the Margin: Henry Savile and Joseph Scaliger on Squaring the Circle" in *Scientia in Margine: Études sur les marginalia dans les manuscrits scientifiques du moyen âge à la Renaissance*, ed. D. Jacquart and C. Burnett, pp. 241–59, Geneva: Libraire Droz, 2005.

Robert Goulding, "Histories of Science in Early-Modern Europe," *Journal of the History of Ideas* 67 (2006a), pp. 33–40.

Robert Goulding, "Mathematics and Method: Petrus Ramus and the History of the Sciences," *Journal of the History of Ideas* 67 (2006b), pp. 63–85.

Anthony Grafton, *Defenders of the Text: Traditions of Scholarship in an Age of Science 1450–1800*, Cambridge, MA: Harvard University Press, 1991.

Anthony Grafton, "From Apotheosis to Analysis: Some Late Renaissance Histories of Classical Astronomy" in *History and the Disciplines: The Reclassification of Knowledge*, ed. Donald R. Kelley, pp. 261–76, Rochester, NY: University of Rochester Press, 1997.

J. Harvey, "Architecture in Oxford 1350–1500" in *The History of the University of Oxford*, ed. J. I. Catto and R. Evans, 2, pp. 747–68, Oxford: Oxford University Press, 1992.

J. L. Heiberg, *Litterargeschichtliche Studien über Euklid*, Leipzig: B. G. Teubner, 1882.

S. K. Heninger, *Touches of sweet harmony: Pythagorean cosmology and Renaissance poetics*, San Marino, CA: Huntington Library, 1974.

J. R. L. Highfield and G. H. Martin, *A History of Merton College, Oxford*, Oxford: Oxford University Press, 1997.

Jens Høyrup, "Archimedism, not Platonism: on a Malleable Ideology of Renaissance Mathematicians (1400 to 1600), and on its Role in the Formation of Seventeenth-Century Philosophies of Science" in *Archimede. Mito Tradizione Scienza*, ed. Corrado Dollo, *Biblioteca di Nuncius: Studi e testi* IV, pp. 81–110, Florence: Leo S. Olschki, 1992.

Nicholas Jardine and Alain Segonds, "A Challenge to the Reader: Ramus on *Astrologia* without Hypotheses" in *The Influence of Petrus Ramus: Studies in Sixteenth and Seventeenth Century Philosophy and Science*, ed. Mordechai Feingold, Joseph S. Freedman, and Wolfgang Rother, *Schwabe Philosophica* 1, pp. 248–66, Basel: Schwabe, 2001.

R. W. Jeffery, "History of the College: 1547–1603" in *Brasenose College Quatercentenary Monographs, Oxford Historical Society Publications* 53 (monograph 10), Oxford: Clarendon, 1909.

G. Lloyd Jones, "Introduction" in Johann Reuchlin, *On the Art of the Kabbalah*, ed. and tr. Martin and Sarah Goodman, pp. 7–32, New York: Abaris Books, 1983.

Christiane Joost-Gaugier, *Measuring Heaven: Pythagoras and His Influence on Thought and Art in Antiquity and the Middle Ages*, Ithaca NY: Cornell University Press, 2006.

Wilbur R. Knorr, *The Ancient Tradition of Geometric Problems*, New York: Dover, 1986.

Wilbur R. Knorr, *Pseudo-Euclidean Reflections in Ancient Optics: A Re-Examination of Textual Issues Pertaining to the Euclidean* Optica *and* Catoptrica, *Physis* 31 (1994), pp. 1–45.

Jill Kraye, " 'Ethnicorum omnium sanctissimum': Marcus Aurelius and his *Meditations* from Xylander to Diderot" in *Humanism and Early Modern Philosophy*, ed. Jill Kraye and Martin W. F. Stone, pp. 107–34, London: Routledge, 2000.

Christopher R. Ligota, "Annius of Viterbo and Historical Method," *Journal of the Warburg and Courtauld Institutes* 50 (1987), pp. 44–56.

François Loget, "La Ramée critique d'Euclide. Sur le *Prooemium mathematicum* (1567)," *Archives internationales d'histoire des sciences* 54/153 (2004), pp. 3–28.

Cora E. Lutz, "Remigius' Ideas on the Origin of the Seven Liberal Arts," *Medievalia et Humanistica* 10 (1956), pp. 31–49.

C. E. Mallet, *A History of the University of Oxford*, London: Methuen, 1924 (in 3 vols).

Jean-Claude Margolin, "L'Enseignement des mathématiques en France (1540–70): Charles de Bovelles, Fine, Peletier, Ramus" in *French Renaissance Studies 1540–1570: Humanism and the Encyclopedia*, ed. Peter Sharratt, pp. 109–155, Edinburgh: Edinburgh University Press, 1976.

James McConica, "Humanism and Aristotle in Tudor Oxford," *The English Historical Review* 94 (1979), pp. 291–317.

Robert McNulty,"Bruno at Oxford," *Renaissance News* 13 (1960), pp. 300–05.

Kees Meerhoff, "Beauty and the Beast: Nature, Logic and Literature in Ramus" in *The Influence of Petrus Ramus: Studies in Sixteenth and Seventeenth-Century Philosophy and Science*, ed. Mordechai Feingold, Joseph S. Freedman, and Wolfgang Rother, *Schwabe Philosophica* 1, pp. 200–14, Basel: Schwabe, 2001.

Charlotte Methuen, "The Role of the Heavens in the Thought of Philip Melanchthon," *Journal of the History of Ideas* 57 (1996), pp. 385–403.

Arnaldo Momigliano, *Ancient Wisdom: the limits of Hellenization*, Cambridge: Cambridge University Press, 1975.

Marian A. Moore, "A Letter of Philip Melanchthon to the Reader," *Isis* 50 (1959), pp. 145–50.

Reviel Netz, "The Aesthetics of Mathematics: A Study" in *Visualization, Explanation and Reasoning Styles in Mathematics*, ed. Paolo Mancosu, Klaus Jørgensen, and Stig Andur Pedersen, *Synthese Library* 327, pp. 251–93, Dordrecht: Springer, 2005.

Walter J. Ong, *Ramus, Method, and the Decay of Dialogue: From the Art of Discourse to the Art of Reason*, Cambridge, MA: Harvard University Press, 1958a.

Walter J. Ong, *Ramus and Talon Inventory*, Cambridge, MA: Harvard University Press, 1958b.

Walter J. Ong, "Christianus Ursitius and Ramus' New Mathematics," *Bibliothèque d'humanisme et renaissance* 36 (1974), pp. 603–10.

Anna de Pace, *Le matematiche e il mondo: ricerche su un dibattito nella seconda metà del Cinquecento*, Milan: Francoangeli, 1993.

Isabelle Pantin, "Ramus et l'enseignement des mathématiques" in *Ramus et l'Université*, ed. Kees Meerhoff and Michel Magnien, *Cahiers V. L. Saulnier* 21, pp. 71–86, Paris: Presses de l'École normale supérieur, 2004.

W. A. Pantin, *Oxford Life in Oxford Archives*, Oxford: Clarendon, 1972.

Olaf Pedersen, "The Origins of the *Theorica Planetarum*," *Journal for the History of Astronomy* 12 (1981), pp. 113–23.

Jeanne Peiffer, "France" in *Writing the History of Mathematics: Its Historical Development*, ed. Joseph W. Dauben and Christoph J. Scriba, *Science Networks* 27, pp. 3–43 (ch. 1), Basel: Birkhäuser, 2002.

Peter Pesic, "Secrets, Symbols and Systems: Parallels Between Cryptanalysis and Algebra, 1580–1700," *Isis* 88 (1997), pp. 674–92.

Heinrich F. Plett, *Rhetoric and Renaissance Culture*, Berlin: Walter de Gruyter, 2004.

Charles Plummer, *Elizabethan Oxford; Reprints of Rare Tracts*, *Oxford Historical Society* 8, Oxford: Clarendon Press, 1887.

Gianna Pomata and Nancy G. Siraisi eds, *Historia: Empiricism and Erudition in Early Modern Europe*, Cambridge, MA: MIT Press, 2005a.

Gianna Pomata, "*Praxis historialis*: The uses of *historia* in early modern medicine" in *Historia: Empiricism and Erudition in Early Modern Europe*, ed. Gianna Pomata and Nancy G. Siraisi, pp. 105–46, Cambridge, MA: MIT Press, 2005b.

Nicholas Popper, " 'Abraham, Planter of Mathematics': Histories of Mathematics and Astrology in Early Modern Europe," *Journal of the History of Ideas* 67 (2006), pp. 87–106.

Stephen Pumfrey and Frances Dawbarn, "Science and Patronage in England, 1570–1625: A Preliminary Study," *History of Science* 42 (2004), pp. 137–88.

Timothy J. Reiss, *Knowledge, Discovery and Imagination in Early Modern Europe: The Rise of Aesthetic Rationalism*, *Cambridge Studies in Renaissance Literature and Culture* 15, Cambridge: Cambridge University Press, 1997.

Timothy J. Reiss, "From Trivium to Quadrivium: Ramus, Method and Mathematical Technology" in *The Renaissance Computer: Knowledge Technology in the First Age of Print*, ed. Neil Rhodes and Jonathan Sawday, pp. 45–58, London and New York: Routledge, 2000.

Christoph Riedweg, *Pythagoras: His Life, Teaching and Influence*, tr. Steven Rendall, Ithaca and London: Cornell University Press, 2005.

Paul Lawrence Rose, *The Italian Renaissance of Mathematics: Studies on Humanists and Mathematicians from Petrarch to Galileo*, Geneva: Librairie Droz, 1975.

Paul Lawrence Rose, "Bartolomeo Zamberti's Funeral Oration" in *Cultural Aspects of the Italian Renaissance: Essays in Honor of Paul Oskar Kristeller*, ed. Cecil H. Clough, pp. 299–310, Manchester: Manchester University Press, 1976.

Paul Lawrence Rose, "Erasmians and Mathematicians at Cambridge in the Early Sixteenth Century," *Sixteenth Century Journal* 8 (1977), pp. 46–59.

Remigio Sabbadini, *La scuola e gli studi di Guarino Guarini Veronese*, Catania: Francesco Galati, 1896. Reprinted in facsimile in Sabbadini, *Guariniana*, Turin: Bottega d'Erasmo, 1964.

Christoph J. Scriba, Menso Folkerts, and Hans Wussing, "Germany" in *Writing the History of Mathematics: Its Historical Development*, ed. Joseph W. Dauben and Christoph J. Scriba, *Science Networks* 27, pp. 109–49 (ch. 5), Basel: Birkhäuser, 2002.

Peter Sharratt, "La Ramée's early mathematical teaching," *Bibliothèque d'Humanisme et Renaissance* 28 (1966), pp. 605–14.

Peter Sharratt, "The Present State of Studies on Ramus," *Studi francesi* 16, pp. 201–13 (1972).

Peter Sharratt, "Nicolaus Nancelius, *Petri Rami vita*. Edited with an English translation," *Humanistica Lovaniensia* 24 (1975), pp. 161–277.

Peter Sharratt, "Recent Work on Peter Ramus (1970–1986)," *Rhetorica* 5 (1987), pp. 7–58.

Peter Sharratt, "Ramus 2000," *Rhetorica* 18 (2000), pp. 399–455.

James Veazie Skalnik, *Ramus and Reform: University and Church at the End of the Renaissance*, *Sixteenth Century Essays and Studies* 60, Kirksville, MO: Truman State University Press, 2002.

Roy Sorensen, *A Brief History of the Paradox*, Oxford: Oxford University Press, 2003.

Walter E. Stephens, "The Etruscans and Ancient Theology in Annius of Viterbo" in *Umanesimo a Roma nel Quattrocento*, ed. Paolo Brezzi and Maristella de Panizza Lorch, pp. 309–22, Rome: Istituto di Studi Romani, 1984.

Noel M. Swerdlow, "Science and Humanism in the Renaissance: Regiomontanus's Oration on the Dignity and Utility of the Mathematical Sciences" in *World Changes: Thomas Kuhn and the Nature of Science*, ed. Paul Horwich, pp. 131–68, Cambridge, MA: MIT Press, 1993a.

Noel M. Swerdlow, "Montucla's Legacy:The History of the Exact Sciences," *Journal of the History of Ideas* 54 (1993b), pp. 299–328.

E. G. R. Taylor, *The Mathematical Practitioners of Tudor and Stuart England*, Cambridge: Cambridge University Press, 1954.

Lynn Thorndike, *The Sphere of Sacrobosco and Its Commentators*, Chicago: University of Chicago Press, 1949.

Charles Trinkaus, "A Humanist's Image of Humanism: The Inaugural Orations of Bartolommeo della Fonte," *Studies in the Renaissance* 7 (1960), pp. 90–147.

Brian Vickers, "Philosophy and humanistic disciplines: Rhetoric and poetics" in *Cambridge History of Renaissance Philosophy*, ed. Charles B. Schmitt, Eckhard Kessler, and Jill Kraye, pp. 715–45, Cambridge: Cambridge University Press, 1988.

Charles Waddington, *Ramus: Sa vie, ses écrits et ses opinions*, Paris: C. Meyrueis, 1855.

Craig Walton, "Ramus and Socrates," *Proceedings of the American Philosophical Society* 114 (1970), pp. 119–39.

Samuel Wheeler, "Megarian Paradoxes as Eleatic Arguments," *American Philosophical Quarterly* 20.3 (1983), pp. 287–95.

Frances Yates, "Giordano Bruno's Conflict with Oxford," *Journal of the Warburg Institute* 2 (1938–9), pp. 227–42.

# Index

The manufacturer's authorised representative in the EU is Springer Nature Customer Service Centre GmbH, Europaplatz 3, 69115 Heidelberg, Germany. If you have any concerns regarding our products, please contact ProductSafety@springernature.com

Printed and bound by CPI Group (UK) Ltd, Croydon, CR0 4YY
22/09/2026
02221058-0011